A Critical History of Schizophrenia

Kieran McNally
Adjunct Lecturer in Psychology,
University College Dublin, Ireland

palgrave
macmillan

First published 2016 by
PALGRAVE MACMILLAN

Palgrave Macmillan in the UK is an imprint of Macmillan Publishers Limited, registered in England, company number 785998, of Houndmills, Basingstoke, Hampshire RG21 6XS.

Palgrave Macmillan in the US is a division of St Martin's Press LLC, 175 Fifth Avenue, New York, NY 10010.

Palgrave Macmillan is the global academic imprint of the above companies and has companies and representatives throughout the world.

Palgrave® and Macmillan® are registered trademarks in the United States, the United Kingdom, Europe and other countries.

ISBN 978–1–137–45680–9

This book is printed on paper suitable for recycling and made from fully managed and sustained forest sources. Logging, pulping and manufacturing processes are expected to conform to the environmental regulations of the country of origin.

A catalogue record for this book is available from the British Library.

Library of Congress Cataloging-in-Publication Data
McNally, Kieran.
A critical history of schizophrenia / Kieran McNally, Adjunct Lecturer in Psychology, University College Dublin, Ireland.
pages cm.—(Palgrave studies in the theory and history of psychology)
ISBN 978–1–137–45680–9 (hardback)
1. Schizophrenia—History. I. Title.
RC514.M42 2015
616.89'8—dc23 2015028739

Typeset by MPS Limited, Chennai, India.

Mesdames et Messieurs, je serais hereux si j'avais réussi à vous montrer que la schizophrénie n'est pas un concept purement théorique et illusoire

Eugen Bleuler, 1926, p. 17

Contents

Acknowledgements

I first developed an interest in the concept of schizophrenia while working at the Institute of Psychiatry in London under the auspices of the effervescent Jeffery Gray, whose lasting impression on me was his compulsive desire to question. It was my additional fortune to later meet Adrian Brock, my thesis supervisor at University College Dublin (UCD), who patiently taught me how a historical approach could be applied to understanding schizophrenia. For both encounters I am extremely grateful.

As academia is a collaborative process many people helped this book emerge in numerous ways. Much of this book, for example, saw light in earlier publications in journals whose peer review process greatly enhanced the book. As such, I owe a great deal of thanks to the insightful comments of editorial teams and numerous anonymous peer reviewers at journals such as *History of Psychiatry*, *History of Psychology*, *Theory and Psychology*, *The Psychologist*, and the *Journal of the History of the Behavioral Sciences*. I also received much assistance from staff at UCD and from various attendees at conferences, such as the annual gathering of the British Psychological Society's History and Philosophy of Psychology Section and the European Conference on Schizophrenia. Conservatively, I put the number of contributors at close to 100 people. There are too many to name, but in the early days I recall that Geoff Burn, Alan Collins, and Jon Sutton gave particular encouragement that the book was much needed. Since then, I have drawn particular inspiration from the encyclopaedic Graham Richards and the seemingly omniscient Richard Noll.

I would also like to acknowledge funding from the IRCHSS (Irish Research Council) and from the Wellcome Trust. And, of course, this book would also not have been possible without the help of the staff at Palgrave Macmillan and Jack Martin, the series editor, as well as Jayne MacArthur, copyeditor.

Last, but not least, this book would not have happened without support from my close friends, beloved parents, and family. I trust you know your names, although Hugh (9) and Sadhbh (6) would doubtless like to see theirs in print. Kids, I am now finished this 'boring old history of skips-o-phrenia'. We can go outside and play in the garden.

Kieran McNally

Introduction

In the early 1960s, two boys diagnosed with schizophrenia reported seeing green monsters. They were hallucinating. Psychologist Leonard Cobrinik reported verbatim some of the boys' words:

> They [the boys] said, 'You are cheating us out of time. You are trying to make us crazy, make us sad, make us do things'. They tended to identify with each other against the examiners. Later one said, 'you were making us see green monsters but we wouldn't tell you' (Bender et al., 1966, p. 485).

The boys were not exaggerating or lying. Around 1961, they had been placed on daily experimental doses of LSD at the Children's Unit in Creedmoor State Hospital, New York. Reports of the experiment were written up in the publication *Biological Treatment of Mental Illness*. The boys had been on the drug for over one year.

Interest in the subject of hallucinogenic substances and schizophrenia was not idiosyncratic for its time. The less conventional R. D. Laing for example, would also use LSD in his therapeutic regime. Nor was it new. Already in 1926 Heinrich Klüver had suggested in the *American Journal of Psychology* that mescal might help elucidate phenomena in schizophrenia (Klüver, 1926). In the children's case, LSD was thought by the experimenters (psychologists and psychiatrists) to work on the central nervous system and the autonomic functions. All the experimenters needed to do, in order to self-justify the usage of LSD, was to produce a definition of childhood schizophrenia—a definition that could in some way be linked to the actions of the drug, supplied by Sandoz. This they easily managed to do:

> Our definition of this condition is a disorder in maturation characterized by an embryonic primitive plasticity *in all areas of integrative brain functioning* from which behavior subsequently arises. This includes all autonomic functions ... (Bender et al., 1966, p. 464, my emphasis).

As such, having covered 'all areas of integrative brain functioning' it was 'hoped' that the drug might prove, 'somewhat specific in modifying the basic processes as well as the secondary symptoms' (Bender et al., 1966, p. 464). Theorisation complete; bring on the monsters.

Sometimes it seemed as if the monsters wore white coats. In 1939, *Time* magazine called the metrazol treatment of schizophrenia *medieval* as it noted:

> So horrible are the artificial epileptic fits forced by metrazol that practically no patients ever willingly submit. Common symptoms are a 'flash of blinding light', an 'aura of terror'. One patient described the treatment as death 'by the electric chair'. Another asked piteously: 'Doctor, is there any cure for this treatment?' More serious than this subjective terror are dislocations of the jaw, tiny compression fractures of the spine, which occurred to metrazol patients in over 40% of one series of cases. During their violent convulsions, patients arch their backs with such force that sometimes they literally crush their vertebrae (*Time*, 1939, p. 7).

Elsewhere, there was a recognised confusion with the term childhood schizophrenia or, as it was known in French, *schizophrénie infantile*. Nevertheless, others would experimentally tear through such brains with psychosurgery, leucotomy, and topectomy (Heuyer, 1974). Let's pause there. For as uncomfortable as kids on LSD and so on are, the disturbing 'treatment' by twentieth century health professionals (experimental or otherwise) does not particularly surprise us. Contentious treatment of patients in the twentieth century is now a dominant historical narrative. And for many, it is now impossible to imagine the history of schizophrenia in its entirety without thinking of variations of such behaviour.

At times this examination of schizophrenia does touch on 'treatments' that no history of caring can negate (and there is one). However, it is not focused on such troubling occurrences. Nor does it focus on the starvation in French and Greek asylums during the Second World War. And it is not about the mass extermination in Nazi Germany and elsewhere of those diagnosed with schizophrenia (von Bueltzingsloewen, 2007). It might have been. As others have shown, these are all important and unquestionable ways of helping us understand twentieth-century schizophrenia. But it is not. Instead, this history seeks to contribute in a different way. It turns its attention and looks specifically at something less likely to provoke an immediate emotional reaction. And in doing so it seeks to enhance our collective understanding of 'schizophrenia', as found in a corpus of works by Noll (2011), Howells (1991), Gottesman (1991), Gelman (1999), Gelinas (1977), Boyle (2001), and many others that have critically informed this book. Moreover, it does so in a narrow rather than broad sense. It simply restricts its investigation to an examination of an intellectual and social abstraction emerging from within and internal to the behavioural professions: the *concept* of schizophrenia itself.

Schizophrenia

The goal of this book is to increase the historical understanding of the concept of schizophrenia in the twentieth century.[1] However, to do so, it's best to start with some broad brushstrokes before getting into finer detail. So to begin with, let us observe that the concept of schizophrenia was originally formulated by Swiss Psychiatrist Eugen Bleuler, in 1908. Drawing heavily on Emil Kraepelin's dementia praecox, Bleuler used the term to reference a hypothesised group of diseases or psychoses—one in which the 'splitting of the psychic functions is the outstanding symptom of this whole group' (Bleuler, 1908, p. 436).

Of the group itself, Bleuler presented many cases that were often fascinating, sensational, and in need of explanation: a woman who claimed she was a shark, a woman who believed her foster daughter was Snow White, and people who claimed they were flayed and burnt during the night. There were individuals who claimed that their bowels had been torn out, who had been threatened by Judas Iscariot, who masturbated openly, and who claimed that their bath water had been poisoned. There were patients who heard voices and who could smell corpses. There were cases that might pluck out an eye, sow stockings on rugs, and who found potatoes evil. There were many who experienced sexual hallucinations and erotic delusions (Bleuler, 1911/1952). In doing so, Bleuler sketched a variety of often insightful symptoms, sometimes present, sometimes not, such as loosening of associations, non sequitar thinking, autism, perseveration, echopraxia (imitation of movement), echolalia (repetition of words), blocking of thought, ambivalence, hallucinations, delusions, paramnesia (distorted memory), motor symptoms, stupor, mannerisms, negativism, command automatism, impulsiveness, melancholia, euphoria, flight of ideas, twilight states (projection of a personal world owing to anomalies of consciousness), *benommenheit* (see later), dipsomania, fugue states, and so forth (Bleuler, 1911/1952).

By 2004, almost a century later, taking stock of decades of research, the American Psychiatric Association (APA), would produce its version of schizophrenia in its latest *Diagnostic and Statistical Manual of Mental Disorders* (DSM). It would surmise that schizophrenia was a disorder that lasted for at least six months and that usually struck adults in their twenties. It included at least one month of active phase symptoms, that is, two or more of the following: delusions, hallucinations, disorganised speech, grossly disorganised or catatonic behaviour (psychomotor disturbances). It also included 'negative symptoms' such as affective flattening, alogia, and avolition (American Psychiatric Association, 2004). Hallucinations and delusions were much less dominant than they appear in the movies.

In any single description of schizophrenia, current psychiatric texts seldom list large numbers of symptoms or signs. Yet for all that, hundreds of supposed and theorised indicators of schizophrenia can be found scattered throughout twentieth-century literature. Without leaving the letter A, that gave us alogia and avolition above, we meet many obscure and now seldom-cited symptoms. Examples include: aporia, analgesia, aboulia, anorexia, asthenia, amenorrhoea, and apophany. But we also meet the more frequently cited ambivalence, automatism, aloofness, anhedonia, apathy, ataxia, and anxiety. As well as: associability, aggression, autism, attention deficits, autoeroticism, and auditory hallucinations. And that's just symptoms beginning with A.

The quantity in itself was not without precedent. In theorising and retheorising dementia praecox, for example, Kraepelin himself had at one point sketched 53 symptoms in 68 pages (Decker, 2013). Bleuler, as we can see, had followed suit. And the conceptualisation of schizophrenia had continued in this vein. Yet, collectively, this heterogeneous constellation of twentieth-century symptoms resisted easy summation. This fact did not go unnoticed in the literature. In 1977 alone we find a litany of complaints. Hogarty observed, 'One is immediately struck by the diversity of symptoms presumed to be critical to the diagnosis of schizophrenia' (1977, p. 588). Holzman noted, 'Its dizzying array of symptoms and its mercurial appearance baffle systematic observers' (1977, p. 588). And Greenberg complained of schizophrenia's evolution 'into a catch all tag for numerous combinations of symptoms' (1977, p. 28).

As is now well recognised, the concept of schizophrenia was not static (Berrios et al., 2003), and the singularity of the word should not mislead. Its meaning changed many times during its history. Some change in a scientific or medical concept is to be expected. In hindsight, however, the extent to which schizophrenia changed over the twentieth century was breath-taking. Beginning with a handful of subtypes, a mere *portion* of the competing variations of schizophrenia devised by researchers makes for a long list. It included acute schizophrenia, ambulatory schizophrenia, acute schizophrenic reaction, and schizophrenic reaction acute undifferentiated type. To that we can add schizophreniform states or psychosis, synchronous–syntonic schizophrenia, asynchronous–asyntonic schizophrenia, and catastrophic schizophrenia. We can also add catatonic parergasia, catatonic–hebephrenic schizophrenia, chronic undifferentiated schizophrenia, and constitutional schizophrenia. It also included cycloid or episodic schizophrenic psychosis, borderline schizophrenia, dementia praecox, and defect schizophrenia. And it included dementia paranoides, dementia praecoxisma, and dementia precox parkinsonoides. There was also heboidophrenia, hebephrenia, five–day schizophrenia, and three–day schizophrenia. And let's not forget slow-flow schizophrenia, incipient schizophrenia, and confusional schizophrenia.

Even the smallest change in the description or interpretation of the concept of schizophrenia could have profound consequences for the contents of schizophrenia classification. The DSM-III, for example, had whimsically included the provision that onset be before the age of 45. This rendered someone unable to be diagnosed after their forty-fifth birthday. But in the revised 'Text Revision' DSM-III-R this requirement was dropped without much fuss (American Psychiatric Association, 1987). (Subsequent researchers would propose *late-onset schizophrenia* as occurring after the age of 45 and *very late-onset schizophrenia* as occurring after the age of 65). DSM-III-R was a supposed nonsignificant text revision. Nevertheless, a previously abandoned childhood-onset schizophrenia had also crept back into what Arieti (1955) once called the 'zoological garden'.

The complexity and abstraction did not stop there. In a feast of splitting, lumping, and synonymising, more subtypes that graced the literature included juvenile schizophrenia, late schizophrenia, and latent schizophrenia. Others included les formes frustes de la schizophrénie, larval schizophrenia, late paraphrenia, paranoia, late indeterminate schizophrenia, and nuclear schizophrenia. Others again included non-praecox catatonia, postemotive schizophrenia, postpartum schizophrenic psychoses, pseudoneurotic schizophrenia, and process schizophrenia. We also find postinfluenzal schizophrenia, schizomania, schizonévrose, schizocaria, schizomanie a forme imaginative, schizonoïa, schizophasia, schizophrenia mitis, schizophrenia deliriosa, and childhood schizophrenia. We further find pseudoschizophrenia, pseudopsychopathic schizophrenia, simple schizophrenia, schizophrenia simplex, schizophrenia restzustand, schizothymia, and senile schizophrenia. And let's close with situational schizophrenia, schizophrenoses, true schizophrenia, transient schizophrenia, unsystematic schizophrenias, and the unmistakable schizophrenias.

Small wonder then that the *Schizophrenia Bulletin's* editor remarked that all too often the patient had become a depersonalized example of a classification scheme (Anon, 1970a). Accompanying this incessant generation of subtypes, which often saw patients reshuffled through multiple categories over time, was a similar flux concerning the personality of the patient. In 1931, Lewis and Blanchard, for instance, showed a personality type table. This outlined three personality types: the 'cyclothyme', the 'hyperaesthetic schizothym', and the 'anaesthetic schizothym'. They also discuss the, 'cold schizothym of Kretschmer [sic]' (Lewis and Blanchard, 1931, p. 484). The shut-in personality described by Hoch was another type that graced the literature (Noll, 2015). Lundholm would speak of allocentric and egocentric personality traits to theorise conduct types for 'extroactive types of schizophrene, the silly hebephrenic and the paranoid' (1932, p. 106), before moving on to tackle what he called the 'royal riddle of the schizoid' (1932, p. 100). The degree to which various types of personality were perceived to be synonymous with the various schizophrenias was dependent on the theory in question. Yet all hovered in the background threatening to complicate schizophrenia's conceptualisation at any time.

During this century, hundreds of abnormalities had been reported in *groups* of individuals diagnosed with schizophrenia. Almost no organ or brain region had been left unimplicated at one stage or another. And, by 1983, faith still existed that rational neurobiological data would 'allow us to solve the puzzle of schizophrenia' (Nicol and Gottesman, p. 403). Yet by official reckoning things were less rosy. A summation of the twentieth-century literature revealed that no laboratory findings had been identified as diagnostic of schizophrenia per se (American Psychiatric Association, 2004). Indeed, 1977 comments by Buchsbaum still haunted the literature. The path to understanding schizophrenia was 'littered with hundreds of discarded physiological and biochemical findings' (Buchsbaum, 1977, p. 7). Investigations yielded 'amazingly wide individual variation' among people diagnosed with schizophrenia (ibid). And consequently, for some, 'the range of defects observed had been too great and too variable from study to study to permit much coherent theoretical interpretation' (Claridge, 1978, p. 186). In a period where intellectual despair sometimes threatened to tear up the concept, poorly designed research confounded the complexity (Morriarty and Massett, 1970). Some seemingly sought facetious refuge in the fact that the diagnosis of normality was 'even less reliable' (Gunderson et al., 1974, p. 32).

A return to cannonical documents did not necessarily deliver clarity. As early as 1925, for example, Sullivan would complain—with partial justification—that Bleuler and Kraepelin had developed their concepts on cases that had already developed beyond help. Sullivan further complained that books such as Bleuler's *Textbook of Psychiatry* 'came illustrated with specimens occasionally of the side-show variety' (1962, p. 32) (an implicit convention in much psychiatric writing). Similarly, many of Kraepelin's patients displayed epilepsy (1913/1919, p. 116), while symptom descriptions at times included specious declarations that 'women put matrimonial advertisements into newspapers' (1913/1919, p. 96), or that women felt that 'their father, their clergyman has abused them ...' (1913/1919, p. 30). Critics looking backward could also note that for all Bleuler's insight, the 'schizophrenic' ultimately remained stranger to him than the birds in his garden (Laing, 1960/1990).

In 1977, Trotter summed up the prevailing confusion by noting, 'Schizophrenia is like a labyrinthine maze through which researchers crawl, searching for a way out. ... The fact that there is evidence to support almost every theory of schizophrenia is what makes it so confusing' (1977, p. 394). In 1978 Wilson would note that, 'Like witchcraft, schizophrenia is "real" but like Macbeth's dagger it cannot be grasped' (1978, p. 92). And by 1988, *Nature* would publish contradictory genetic studies side by side. In subsequent commentary on these and other studies, Dixon would allude to the 'increasing conviction of the aetiological heterogeneity of schizophrenia' (1989, p. 265). Understandably then, schizophrenia was sometimes also understood by what it was not. It was a popular misconception, texts cautioned, that schizophrenia meant a split personality, or multiple personality (Klerman, 1978; Sutherland, 1976). The schizophrenic, it seemed, had become 'psychiatry's "quintessential other" the patient whose very essence is "incomprehensibility" itself' (Sass, 1987, p. 4).

As is already clear, twentieth-century schizophrenia was not a simple scientific concept. As Ruth Leys (2000) has noted of trauma, it refuses to bend to a simplistic idea of a linear, if interrupted, development. So if we ask 'What did the twentieth century really understand by the concept of *schizophrenia*?', we can already recognise that this is an open question easier asked than answered. Yet ask it we must. Schizophrenia was psychiatry's arch disorder. And it still is. We need to try and understand how such an apparent intellectual quagmire unfolded. Such a history not only facilitates our understanding of twentieth century ideas of madness (a concept in its own right), but it also impacts on our current understanding of contemporary schizophrenia, the signature concept of modern psychiatry. To understand schizophrenia without understanding its history is tantamount to understanding nothing at all.

A Brief Outline

This work preferentially examines schizophrenia though engaging some of the very spaces and relationships in which twentieth-century madness was declared visible. As Danziger and Daston have argued, scientific realism must take the historicity of scientific objects seriously (Danziger, 1993; Daston, 2000). Equally, it is now apparent that the perceived unity of a disorder is partially bonded by the narratives with which it is represented (Young, 1995). In this vein, the work attempts a critical examination of the concept's changing definitions, symptoms, and classification. It also examines the metaphor of the split personality. And, more generally, the work alerts us to the very language through which the concept has been formulated. Such an approach draws inspiration from Michel Foucault's study of order and representation in *The Order of Things*. Therein, Foucault draws attention to both the possibilities and limits to representation (Foucault, 1966/2006). It is also partially inspired by Nikolas Rose's identification of the utility of exploring the language and grammar of explanatory systems. That is, through rhetoric, metaphors, analogies, logics, and so on (Rose, 1999). Naive realism must be set aside.

In looking at the concept in the above manner, this history further seeks to understand schizophrenia's contradictions and how it came to be socially negotiated, maintained, and transformed. And in doing so it resurrects the theorisation voiced by the concept's architects and the responses of critical voices that called their ideas into question. Such an investigation is further based upon a simple belief. By first understanding the historical representation of our concepts, it becomes possible to free ourselves from their limitations. After that we can begin to catalogue and contextualise empirical findings, and so on, in a more useful way. Without such knowledge, historical research in the history of psychiatry will be impoverished, as indeed might day-to-day empirical science. For such reasons, preceding its conclusions, the book is roughly structured as follows.

Chapter 1: 'Schizoidia: The Lexicon'

The history opens with an overview of how psychiatry viewed the role of language in conceptualising schizophrenia. The work details how aspects of language came to be seen as problematic in the conceptualisation of schizophrenia. It further explores how the production of many technical descriptives increasingly coagulated and choked the twentieth-century research literature. Some examples might have included Mapother's 'Knights move' in thought associations (Wilson, 1951). Others might include 'dementia schizophrenica', and 'catatonin'—a substance that supposedly produced catatonic symptoms (Cameron, 1935). [For details of Cameron's extensive torture of patients, see Wyden (1998).] However, such a lexicon is too vast to include in its entirety. Instead we acknowledge and explore its existence through a restricted examination of variants of the word schizophrenia. Such an overview usefully foreshadows and contextualises much of the content in later chapters.

Chapter 2: 'The Split Personality'

This history next looks at schizophrenia through metaphor, an essential constitutive element of modern science (Micale, 1995). It explores how schizophrenia came to be conceptualised as a split personality. It also explores to a lesser extent the idea of Jekyll and Hyde. As will be made clear, the stigmatising idea of schizophrenia as a 'split personality' cannot be dismissed as an idiosyncratic misinterpretation by the general public. Rather, to a large extent, it was generated, maintained, and reinforced from within the culture of the psychological professions until it was no longer useful. As such, twentieth-century schizophrenia research emerges as being complicit in the creation of a stigmatising force: one that remains to this day.

Chapter 3: 'Definitions of Schizophrenia'

Part of this book is also about the historical attempts to define schizophrenia. For example, when Mednick theorised in 1958 that schizophrenia was a learned evasion of life (Trotter, 1977), or when, in the Soviet Union, Gurevich and Sereiskii's 1946 *Textbook of Psychiatry* defined schizophrenia as a formal disorder of the psyche that led to a qualitative degradation of the whole personality (Zajicek, 2014). Taken individually, definitions presented an effective façade of knowledge and an illusion of certainty in schizophrenia research. However, to simply see definition as objective facts, the reflection of natural kinds, or the sum of a progressive accumulation of knowledge is self-deceiving. Instead, changing definitions served as social interfaces to the negotiation of the meaning of dementia praecox and schizophrenia. They also served, among other things, as tools of pedagogy and theoretical expression. Equally, definition further depended on expedience and purpose. Ultimately, definition was sidelined in favour of operational approaches in schizophrenia research. For the twentieth century, there was no such thing as a definitive definition of schizophrenia. The concept resisted synopsis and synthesis.

Chapter 4: 'Catatonia: Faces in the Fire'

This chapter examines the 'disappearance' of catatonia. Bleuler included catatonia within his disease concept with deference to Karl Ludwig Kahlbaum, who had coined the term in 1874 (Kahlbaum, 1874/1973). Bleuler noted that, 'More than half of the institutionalized schizophrenics show catatonic symptoms either transitorily or permanently' (1911/1950, p. 180), and as such considered catatonia one of the fundamental subtypes of schizophrenia. Over the next century progress in treating those diagnosed with schizophrenia was modest. Yet by 1987, the APA would state that occurrences of catatonia were *rare* (American Psychiatric Association, 1987). This chapter shows how a subtype whose most notable characteristic increasingly became absence remained tolerated within the twentieth-century conceptualization of schizophrenia. Schizophrenia could tolerate ambiguity.

Chapter 5: 'Chasing the Phantom: Classification'

The conceptualisation of schizophrenia was replete with other subtypes. Catatonia aside, the book does not weigh up each attempt at subtyping—for example Kleist's 'systematic' and 'unsystematic' schizophrenias. Nor does it contrast them with something other. For example, Kleist's work is not compared with Langfeldt's genuine, undoubted, or unmistakable schizophrenias, that Langfeldt divided into two groups: the predominantly endogenously conditioned 'process' schizophrenias, and the atypical schizophreniform states (Langfeldt, 1937, p. 221).[2] Rather, this chapter examines the overall context in which such thinking took place. Schizophrenia taxonomy was constantly fluctuating and re-configured, resulting in the appearance and ostensible disappearance of subtypes. As such, there were distinct echoes of Hoche's 1912 comments with respect to dementia praecox; it often seemed researchers were chasing a phantom (Hoche, 1912).

Chapter 6: 'Myth and Forgetting: Bleuler's Four As'

This chapter explores how a simple twentieth-century mnemonic, the Four As, came to distort Bleuler's complex descriptive pathology. At no stage did Bleuler give precedence to the Four As or describe them in such a fashion. Yet, the Four As emerged as a caricatured representation of Bleuler's schizophrenia that partially distorted the later conceptualization of schizophrenia, masquerading as historical fact. This chapter clarifies the precise relationship of the Four As to Bleuler's thinking. It discusses their emergence and persistence, and draws attention to Bleuler's emphasis of other important symptoms. In doing so, this chapter does not overstate the importance of Bleuler's symptoms. Rather, it further illustrates how a process of historical forgetting and myth making accompanied the conceptualisation of twentieth-century schizophrenia.

Chapter 7: 'Social Prejudice'

Twentieth-century schizophrenia was not immune to racial prejudices and ethnocentric beliefs. It was further intertwined with negative beliefs concerning sexual behaviour, the family, criminality, and political dissent. Indeed, as late as 1957, the *British Medical Journal* could still reference schizophrenia as the most sinister of mental disorders (Anon, 1957a). This is not surprising. Twentieth-century psychiatry had no inherent immunity to the prejudices of the culture of which it was part. Yet this chapter reveals, to an unsettling degree, just how flexible schizophrenia was in its ability to nurture and bend to the prejudices of society.

Chapter 8: 'Contesting Schizophrenia?'

Throughout the twentieth century, criticism of schizophrenia emerged from many sources. The so-called 'antipsychiatry' movement was one source of this doubt. And this chapter will sample some of its conceptual arguments in detail. As we shall see, criticism was not exclusive to antipsychiatry—as many might defensively dismiss. Indeed, as can be seen here (and throughout the book) doubt and objections concerning the concept were not unusual. Frequently, concern came from the very professionals who maintained the concept. In analysing this dissent, we will discover just why schizophrenia's critics considered it to be, 'a semantic Titanic, doomed before it sails—a concept so diffuse as to be unusable in a scientific context' (Bannister, 1971, p. 72).

Chapter 9: 'Manufacturing Consensus in North America'

Finally, we examine how, in the face of raging disagreement and conceptual disorganisation, the APA continually conceptualised and reconceptualised schizophrenia classification. We see how schizophrenia moved from being the theoretical proposition of one psychiatrist, Eugen Bleuler, to being a vast social concept. A concept upheld by a global community of researchers, now dominated by North American psychiatry. It shows how psychiatry, controversially and dubiously, formulated a consensus idea of an objectively knowable schizophrenia through the authority accorded to classification committees.

Caveats

Seeking insight into twentieth-century schizophrenia, this work preferentially explores the history of the *concept* of schizophrenia in its most *abstract* representation in the schizophrenia literature. If the book succeeds it will greatly facilitate our historical understanding of schizophrenia. But the reader should not assume that all the concept's abstractions encountered in this book ever made it into day-to-day practice of clinics and asylums. Many did not. As Noll (2015) has noted, theory should not be confused with practice.

The reader should also note a number of other limitations. First, as others have cautioned, there are limits to applying our own concepts to the past (Leudar and Thomas, 2000). Second, the work sets aside, rather than ignores, psychobiography and the literature dealing with various important themes such as the subjective phenomenology of madness.

Regretfully, it also sets aside issues concerning internationalisation (cf. Brock, 2009), gender, especially within catatonia, and the literature on treatment and broader cultural perspectives. It sets aside the jostling of many other concepts and regretfully brackets out the history of empirical research, leaving them for future work. For the most part, it focuses only on developments within the Anglophone literature where the locus of global psychiatric authority now resides. And although the book does look at the nineteenth century, the study of the concept's prehistory is somewhat curtailed. Reader's looking for more detailed prehistory could start with books that inform this research. Useful starting points include Richards' *Mental Machinery* (1992), Lundbeck's *The Psychiatric Pursuasion* (1994) Goldstein's *Console and Classify* (1990), and Berrios's *The History of Mental Symptoms* (1996). Ditto for Engstrom's *Clinical Psychiatry in Imperial Germany* (2003) and Noll's *American Madness* (2011). This book's focus is also necessarily internalist with all the limitations that implies: principally, it largely focuses on the actual formulation of the concept rather than the broader social conditions that made this possible. Finally, for the reader who wishes to extrapolate these findings to the present, bear in mind that for the most part conceptualisation post-DSM-III is not analysed in this account. The exponential surge in schizophrenia research after this period remains to be analysed by historians. Hence, although it speaks of the twentieth century, this history largely concerns the period 1908–87. But perhaps this is not a bad thing. History always benefits from a little distance.

1
Schizoidia: The Lexicon

To understand the history of any concept we must necessarily interface with the words through which the concept was articulated. Twentieth-century schizophrenia, for all its perceived objectivity, was first and foremost linguistically encoded. Yet it needs to be understood that twentieth-century schizophrenia had a peculiar relationship with words. For most of the twentieth century, an absence of clear biological markers meant that language played a dominant role in schizophrenia conceptualisation. This almost certainly gave rise to certain excesses of vocabulary and a variety of problems that would probably not have existed otherwise.

Let's begin with the 'schizophrenic'. From the beginning of this concept's articulation, people diagnosed with schizophrenia were indistinguishably merged with their disorder. They became referenced as an object known as the 'schizophrenic', or less commonly the cognate 'schizophrene'. On rare occasions such usage could facilitate an intentioned display of affection, as in *The Listener's* 'you dreamy schizophrene' ('Dreamy Schizophrene', 1968). Yet ultimately the results of such objectification were perceived as negative. A person is simply not a disease (imagine being called 'the cancer'). And that's leaving aside the stigma of the split personality (see later).

In a worthwhile, if belated, attempt to rehumanise their object, the APA's DSM-III-R would reject 'schizophrenic' in favour of 'a person with schizophrenia' (American Psychiatric Association, 1987, p. xxiii). There were no longer schizophrenics—only people suffering from schizophrenia. As it happens, the creation of the adjective had not been accidental. Bleuler—doubtless thinking of the adjectival challenges presented by 'dementia praecox'—had deliberately chosen it. For without such a new term, a differential diagnosis 'would be hard to write and even harder to read' (Bleuler,1911/1952, p. 7). Yet in creating the schizophrenic, Bleuler had conveniently embodied a concept that was still merely theoretical. And alongside the catatonic, a new class of person was placed on the psychiatric stage.

Language impacted upon the concept in other ways. By 1968, the APA's DSM-II now interchangeably referred to the concept as schizophrenia (singular), and the 'schizophrenias', a group of disorders (plural). The use of the term 'schizophrenia', explained Bellak and Loeb, had helped to perpetuate a tendency to think of 'this syndrome as a discrete, single, unitary disease' (1969, p. i). In research, it had led to looking for one aetiological or pathogenic factor. In clinical practice, 'the pseudo unity has helped obscure differential diagnosis, prognosis, and treatment; it also confounds and often sadly misleads the general public' (ibid). (In fact, the pseudo unity stemmed from Bleuler's own writing not clinical practice per se.) Bellak's use of 'syndrome' emerged from a literature review of 3200 papers (Noll, 2015). Yet others were less certain, and for them there remained, 'the vexing problem of whether we are dealing with a single disease entity or with a different number of "schizophrenias"' (Brill et al., 1969, p. 110).

Other problematic weaknesses in the articulation of schizophrenia were noted in the twentieth-century literature. And for some, rather than facilitating communication, the language of schizophrenia would come to be seen as disrupting communication. In 1913, Brissot, reporting criticisms by Trénele, would affirm that the invention of the 'misty' concept of schizophrenia encompassed a large number of conditions and that what had occurred was the replacement of a word (dementia praecox) with aetiological pretensions by one with pathological pretensions (Brissot, 1913). In 1971, Bannister claimed the logical utility of the 'already vague' concept or 'omnibus' had not been improved by making it a qualifier, via 'schizoid' or 'schizophreniform' (Bannister, 1971). While in 1975 French psychiatrists argued that using the adjective schizophrenic over the substantive schizophrenia contributed to an abusive expansion of the concept in American psychiatry. This, it was declared, had resulted in a loss of its comprehensibility (Ey et al., 1977). Such complaints hinted at a deep unease with the conceptualisation of schizophrenia and at times with psychiatric language itself. Indeed, the iconoclast Szasz would accuse Bleuler and his followers of having transformed our idea of illness and our vocabulary for describing and defining it. For Szasz they had displaced lesion by language (Roth, 1977).

We need to contextualise such a claim in order to understand it better. As this twentieth-century concept unfolded, a schizophrenia-reifying lexicon developed within a vast and wide-ranging research literature. Some of it was trivial, merely work-a-day jargon and conceptual froth. Yet other aspects of this lexicon took on greater meaning. The lexicon described traits peculiar to 'schizophrenics' and other theoretical suppositions that added legitimacy to the concept. It codified both schizophrenia and the social identity of those who conceptualised it. The magnitude of authority and confidence it projected was enormous. This can be understood by merely examining observations, theoretical presumptions, behaviour, or phenomena labelled with variants stemming from schizophrenia or more often 'schiz'. In itself, the prefix 'schiz' was not unknown prior to the turn of the twentieth century. Stedman's dictionary, for example, included terms such as schizaxon—a neuraxon divided into two branches. It further included schistocephalus (a monster with a cloven head) and schistotrachelus (a monster with a cleft of the neck) (Stedman, 1911). By the end of the twentieth century, however, its variants had proliferated wildly.

From as early as 1910 for example, a person with a mild variant of schizophrenia could be considered 'schizoid' (Bleuler, 1972/1978). This was followed quickly by the usage of 'schizothymia reactiva' to describe a tendency to particular types of psychic splitting with acute onset (Bornstein, 1917). Such a hypothesised psychic disposition or tendency, which could be accentuated with posthypnotic suggestion but differed from hysteria, gave us the 'schizothymic personality'. And later, by extension, it gave the 'schizothymic family'.[1] In 1924, Claude would speak of morbid dream states and dissociation integral to 'schizomanie' or 'états dits schizomaniaques'. In such a scenario patients would flee reality and create their own imaginary worlds. This so-called schizomania was not to be confused with mania (Minkowski, 1927, p. 193). Claude would also speak of the 'schizoses' to embrace all things schizophrenic, although the term later referenced an intermediate group supposedly existing between neurosis and psychoses (Soccaras, 1957). Elsewhere, early twentieth-century Soviet psychiatry would speak of a 'schizoid neurosis' as a normal response to abnormal circumstances (Zajicek, 2014).

In 1925 Lewis discussed the graphic art productions of patients with schizophrenia. In doing so, he used the term 'schistic production' to represent art drawn by schizophrenics. Hence, 'The schistic production is strangely fantastic and unreal or frankly infantile' (Lewis, 1925/1928, p. 367). Schizonoïa, or the 'schizonoiac', argued Laforgue (1927), referenced individuals whose development was disrupted in early emotional relations with their mother. And in the spirit of Fritz

Schulhof, 'schizobulia' came to refer not only to a split of will, but also to a pathological inability to make decisions (Beigel, 1971; Schulhof, 1928). In 1930, Bleuler declared that the degrees of psychopathy in evidence could best be described as a 'schizopathy' (schizopathie) (Bleuler, 1930a).

In 1931 Jacques Lacan and colleagues introduced the word 'schizographie', to describe a special form of 'schizophasia' (*schizophasie*), that is, certain forms of more or less incoherent language (sometimes called 'schizophrenese', which for some, such as Hill (1955), was also a quality of thought). In certain cases such incoherence only manifested itself in written language (Lacan, 1975, p. 365). Schizophasia itself was sometimes used to describe a *regressed* from of language (Bleuler and Claude, 1926/2001, p.54). And elsewhere deviations in syntax and diction could suggest 'schizophrenicity of communication' (Forrest, 1976). Similarly, Bobon's investigations of spontaneous drawings would produce schizoparalexia, schizoparagraphia, and schizoparaphasia (Bobon, 1952, 1967). The term 'schizophrenic surrender' was used to designate the impression of self-abandonment, of surrender, and of acceptance of life at a lower automatic level. This was due to constitutional inadequacy (Campbell, 1943). Harry Stack Sullivan would speak of states of 'schizophrenic perplexity' that made observable the regression of the personality processes. In this vision the patient lived in a world and participated in interpersonal relations, which were dreamlike in varying degrees (Sullivan, 1939/1953).

Such euphonic terms appeared as symbolic incantations in a seemingly unassailable body of knowledge. They are now often entirely redundant. But they represent useful historical markers for anyone trying to understand the history of the concept. It is consequently important that in introducing the concept we remember their historical presence. Not only will this contextualise our initial introductory understanding of schizophrenia as a concept, but it will also further serve as contextual support for the book's later critical attention towards historical attempts to conceptualise schizophrenia in other ways. There are a few more examples worth mentioning briefly.

In 1943, Mira's *Psychiatry in War* would speak of acute active organic processes, which led to the 'schizophrenization' of the individual. By using 'Myokinetic psycho diagnosis' (essentially drawing various lines) the prognosis of the course of schizophrenic syndromes could be found. This involved the persistence or absence of 'schizopraxic' signs in the left hand, corresponding to the deeper layers of the personality (Mira, 1943). Jaspers, could sense a 'schizophrenic atmosphere' in the works of Van Gogh and Hölderlin (Jaspers, 1949/1977). Features of the 'Schizophrenic style' were sometimes found in the works of art produced by schizophrenic patients (Mayer-Grosset et al., 1960). The term 'schizophrenoid psychosis' found favour with Bellak. It distinguished, 'a particularly acute, brief, and benign disorder with schizophrenic symptomatology, but characterised by a dream-like confused state with a prominence of symptoms resembling a hysteria' (Bellak, 1947/1952, p. 447.) 'Schizokinesis' described an inherent conflict between general emotional responses and more adaptive responses. Among other things, this led to the symptom of negativism (Grant, 1953). A regressed portion of the schizophrenic's ego was, for Guntrip (1969), 'the schizoid citadel'.

For Deleuze and Guattari, 'schizoanalysis' was a keyword. It treated, 'the unconscious as an acentered system, in other words, as a machinic network of finite automata (a rhizome)' (2004, p. 19). It arrived, 'at an entirely different state of the unconscious' (ibid). For Johnstone, early life experiences from the first few weeks of life, recalled under methylphenidate, and without which it was theorised that schizophrenia could not occur, were labelled 'schizexperiences' (Anon, 1972a, p. 263). Meehl gave us an inherited neural integrative defect known as 'schizotaxia'. He also gave us the 'schizogene', although 'schizophreniologists' would renounce the possibility of there being a Mendelian disorder at play (Gottesman and Shields, 1976, p. 376). The 'schizophrenic float', referred to the fact that some patients had a peculiar way of walking (Meehl, 1973). The 'schizophrenic smell', reminiscent of stale sweat in unwashed clothes, could be used to distinguish the chronic schizophrenic from 'normals' (Jonas and Jonas, 1975). [On Ludwig Binswanger and *smelling* schizophrenia see Minkowski (1927).] A schizogen or schizomimetic was a drug that produced a state, 'resembling or mimicking naturally occurring psychosis, especially schizophrenia' (Hinsie and Campbell, 1970, p. 633). A disruption in neural circuitry between the cortex and cerebellum would be hypothesised as 'a schizophrenia that is due to a schizencephaly' (Andreason, 1999, p. 782). Yet others would speculate on the existence of a 'schizotoxin' (Gillin et al., 1976), or 'viral schizophrenia' (Anon, 1978a). This would lead to the hunt for a 'schizovirus' (Torrey, 1988). With respect to the latter, Torrey informs us that the term 'schizovirus', along with the term 'schizococcus', were initially whimsically applied to his research for a virus by his colleagues. Even historians of psychiatry became infected, as when, in 1986, Roudinesco could speak of schizophilic theories (Roudinesco, 1986).

The creation, use, and demarcation of terminology constituted a serious attempt to claim new knowledge. Accordingly, the *British Medical Journal* would happily publish letters like Felix Post's nuanced argument 'that we should speak of "complete" rather than of "idiopathic' schizophrenias"' (1963, p.1734). Similarly, Post argued that 'The term "partial" schizophrenia should be substituted for "symptomatic" as well as for "paranoid" schizophrenia' (ibid). Elsewhere, individual articles might also carefully attempt to clarify terms in use. Hemphill, for example, wrote, 'In this paper the term puerperal schizophrenia excludes cases of mixed affective schizophrenia, which appear to belong more to the affective psychoses than the true schizophrenias' (1952, p. 1234). Conversely, and complicating matters, terms could often be considered synonyms. For example, 'Situational schizophrenia' (a hysterical psychosis of wish fulfilling nature) was also considered a form of the schizophrenia precipitated by a severe physical, social, or sexual trauma, known as 'postemotive schizophrenia' (Milici, 1939; Milici and von Salzan, 1938). We will see more of synonymy in relation to schizophrenia and dementia praecox imminently.

The list of incestuously cited terminology was then nigh endless. Hence, the existence of schizophrenia or otherwise was, to paraphrase Dorothy Rowe (1980), as much a battle of words as it was of facts. Recognising this, a frustrated Karl Jaspers, in his magisterial *General Psychopathology*, would complain of pseudo insight through terminology (Jaspers, 1962). [Although in *Strindberg and Van Gogh*, Jaspers himself proposed the possible existence of two dubious categories of schizophrenia. Jaspers wrote: 'Hölderlin and van Gogh represent a type which contrasts sharply with that illustrated by Strindberg and Swedenborg' (1949/1977, p. 194).] Yet for all that, psychiatry did not necessarily despair. The ever-possible discovery of a simple healing chemical or 'magic bullet' might helpfully sweep the whole lexicon into the dustbin of history.

Theoretically, the lexicon was entirely dispensable. Yet while it lasted, this shared hermetic output functioned as an interface to debate and evidence of learning, progress, and expertise. It functioned as currency across diverse networks of people, discourses, and institutional practices. Trainees were initiated into it (necessarily). And where swallowed uncritically, it facilitated, among other things, the ultimate form of professional deception—self-deception. But even when the research in question faded, a fossilised vocabulary left the illusion of an accretion of knowledge. And, at times, it left a false sense of certainty that for some made the validity of schizophrenia seem ever more authoritative.

The schizophrenia lexicon would not function by itself to legitimise the behavioural professions' incorrigible claims to social authority over twentieth-century madness. Other forces were also in play. Nevertheless, for patients and families, all encounters with the conceptualisation of twentieth-century schizophrenia would meet this daunting and incessant articulation. And yet, as we shall see later, for all its seeming authoritativeness, nothing was quite as certain as it seemed.

2
The Split Personality

In the twentieth century, many members of the North American populace came to believe that schizophrenia signalled a 'split personality'. Sometimes even a 'Jekyll and Hyde personality'. By contrast, late twentieth-century students of the mind quickly discovered that this immensely stigmatising belief was *not* the case. Instead, psychiatric textbooks, public campaigns, and psychological course materials cautioned the student of psychology about making such an elementary error. In one way or another, students learned that violence was rare (true). They learned that schizophrenia was commonly misinterpreted by the public as a 'split personality' and that 'the schizophrenic does not suffer from split personality' (Carlson et al., 2004, p. 779). Introductory texts on schizophrenia pretty much left it at that. However, there is a little more to the story behind this divergence between the public and professionals in their understanding of the term schizophrenia. And it makes a useful and necessary point of departure for further easing ourselves into the history of the concept.

The Metaphor of Splitting

The metaphor of splitting can be readily found in various nineteenth-century disciplines, including scientific psychology, philosophy, and literature (Berrios et al., 2003). Most pertinently to schizophrenia, the metaphor took on its most vivid form in Robert Louis Stevenson's 1886 (1994) supernatural horror story *The Strange Case of Dr. Jekyll and Mr. Hyde*. A criticism of Victorian morality, possibly inspired by ergot poisoning, it drew on theological and literary influences concerning humanity's primitive capacity for good and evil. It also drew on conclusions found in Charles Darwin's *Descent of Man* (1871/1981), as well as the work of Herbert Spencer. Similarly, it drew on communication with the French psychiatrist Pierre Janet (Hacking, 1995). Stevenson speculated that 'man will ultimately be known for a mere polity of multifarious, incongruous, and independent denizens' (1886/1994, p. 70).

In the work, Stevenson speaks of 'man's dual nature' (1886/1994, p. 68): 'I now had two characters' (1886/1994, p. 74) and 'My two natures had memory in common but all other faculties were most unequally shared …' (1886/1994, p. 79). Only once does he use the term 'personality', which was then emerging as a conceptual rival to 'character' and as a quality of being somebody (Susman, 1984). That's when he states, 'I shall again and for ever re-endure that hated personality …' (Stevenson, 1886/1994, p. 88). However, Stevenson does not use the words 'split' or 'splitting' anywhere in the text. Nor does he use the terminology 'split personality'. Instead he uses the closely related term 'dissociated': 'in the agonised womb of consciousness, these polar twins should be continuously struggling. How, then, were they dissociated?' (Stevenson, 1886/1994, p. 71). The story became a best seller in Great Britain and America, when it was published in 1886. As such, the idea of the Jekyll and Hyde personality quickly passed into popular mythology.

There is obviously no evidence to suggest that the public at this time conceived of the Jekyll and Hyde personality as schizophrenia. This is simply because the word schizophrenia had yet to be coined. In fact, the Jekyll and Hyde personality would first become bound to the idea of multiple personality—now called dissociative identity disorder. [The history of multiple personality and schizophrenia are not mutually exclusive and has been addressed elsewhere by others such as Ian Hacking (1995).] As such, in 1915, Morton Prince's celebrated multiple Mrs Beauchamp (Prince, 1906) was noted by the *Washington Post* as 'the best example real life has yet afforded of conditions like those which Stevenson imagined for his Dr. Jekyll and Mr. Hyde' (Steiner, 1915, p. 4). As Heinze (2003), in 'Schizophrenia Americana' has also noted, Mrs Beauchamp was also regarded by journalists in 1906 as a case of dual personality. Hence, as it seems clear that at various social levels certain attitudes relevant to splitting and the person pre-existed prior to their formal articulation in schizophrenia research (Berrios et al., 2003; Susman, 1984).[1] As such, what concerns us here now is how—given those receptive conditions—the idea of the 'split personality' became explicitly linked in the popular imagination with that of *schizophrenia*. In passing, we can also improve our understanding of how schizophrenia became linked to the Jekyll and Hyde personality, and to a lesser extent our understanding of the linkage of Jekyll and Hyde with dementia praecox.

Bleuler's Split Personality

Let's return then to Eugen Bleuler, who first coined the word schizophrenia in 1908. Eugen Bleuler was born in the year 1857 in Zollikon, a small town near Zürich in Switzerland. Having studied medicine in Zürich, Bern, and Munich, he worked briefly in Paris. There he was introduced to Benedict-Augustin Morel's (1809–73) theories of degeneration by Valentin Magnan (1835–1916). He also encountered hypnotism through Jean Martin Charcot (1825–93) (Stotz-Ingenlath, 2000). Bleuler then moved to Munich to study neuroanatomy in the laboratory of Bernhard von Gudden from 1884 to 1885 (Hell et al., 2004, p. 34). In 1885, he returned to Zürich as an intern at the Burghölzli asylum. This was followed by a period as the director of a psychiatric nursing clinic at Rheinau from 1886 to 1898. In 1898 he then returned to Burghölzli as director, where, surrounded by dozens of capable investigators (Hell et al., 2001), whose ideas he was very open to, he worked for the rest of his life. It is important to stress that he did not work alone. The history of schizophrenia is not one of autonomous genius. Carl Jung, Sabina Spielrein, Françoise Minkowska, various patients, and Eugen Bleuler's wife Sophie Hedwig Bleuler were but a few of the numerous interesting people who contributed to the concept's formation (Jung's contribution, for example, has been described as pivotal (Shamdasani, 2005)).[2] Perhaps thanks to such contributions, Bleuler's writing was frequently a conduit and point of convergence for many nineteenth-century ideas. Obituaries declared him to have been likeable, paternalistic, and somewhat eccentric.

In 1911, Bleuler published his key text *Dementia Praecox or the Group of Schizophrenias*. Therein he suggested that the name schizophrenia was a useful alternative to Emil Kraepelin's dementia praecox. This was, in part, because he felt that the name dementia praecox (loosely: premature dementia) was misleading with regard to its emphasis on degeneration. Psychiatrists could be heard making the argument 'that the whole concept of dementia praecox must be false' (Bleuler, 1911/1950, p. 8). This was because many catatonics did 'not go on to complete deterioration' (ibid). Bleuler saw his work as an extension of Kraepelin's work on dementia praecox rather than as a replacement of it. Nevertheless, there existed clear differences in the thinking of the two men (for details see later and Hoenig (1995)). Bleuler, in particular, showed a willingness to apply Freudian concepts in mapping the disorder. Hence, in his 1911 preface, he noted the importance 'of the application of Freud's ideas to dementia praecox' (1911/1950, p. 1). Although for all that, his interest in psychoanalytic ideas would become more critical over his career. Bleuler also consolidated his thoughts on schizophrenia in his 1916 (1924) *Textbook of Psychiatry* and elsewhere.

Bleuler's *Dementia Praecox or the Group of Schizophrenias* was not translated into English until 1950 (Bleuler, 1911/1950). Nevertheless, owing to its importance as a psychiatric text, and the prominent influence of early twentieth-century German psychiatry, its ideas would have been in circulation among English-speaking psychiatrists and psychologists. Many were somewhat more practised in German than their modern counterparts. In the work, Bleuler defines schizophrenia as a disease. Not as 'split personality' or 'split mind'. It was a disease 'characterized by a specific type of alteration of thinking, feeling, and relation to the external world' (Bleuler, 1911/1950, p. 9). Bleuler further adds to his definition: 'In every case we are confronted with a more or less clear-cut splitting of the psychic functions' (ibid).

Now, as modern texts sometimes emphasise, there is a clear difference between the splitting of the *psychic functions* and the splitting of the *personality*. Investigating the chapters on accessory symptoms, however, the idea of more than one personality existing within an individual is clearly marked by its author. Bleuler noted, for example, 'In a few cases the "other" personality is marked by the use of different speech and voice ...' (1911/1950, p. 147). Similarly, he stated, 'Thus we have here two different personalities operating side by side, each one may communicate with both' (ibid). Elsewhere Bleuler noted that 'The splitting of the psyche into several souls always leads to the greatest inconsistencies' (1911/1950, p. 129). He also noted that 'the schizophrenic certainly has as many personalities as he has complexes—personalities which are more or less independent of each other' (1911/1950, p. 362).

As noted earlier, the possible sources of inspiration for Bleuler's use of the splitting metaphor are multifarious in this period. But in the case of the latter quote at least, he cites Carl Wernicke's work. His patient in 1900 had consisted 'simultaneously, of a number of different personalities' (1911/1950, p. 361). It should also be noted that Bleuler had absorbed into schizophrenia the dementia praecox of Kraepelin's *Psychiatrie*, a textbook that contains the following passage:

> With dementia praecox in particular this splitting of self-awareness can become very considerable. The sick then speak of foreign powers, enemies, who are in their bodies ... *(Namentlich bei der Dementia praecox kann diese Spaltung des Selbstbewusstseins sehr deutlich werden.*

> *Die Kranken sprechen dann von den fremden Mächten, Feinden, die sich in ihren Körper eingestet haben ...)* (Kraepelin, 1903, p. 236).

[In translation, Kraepelin is also noted as having stated that a dual personality or splitting of self-consciousness occurred in mental disease. And that 'splitting of self-consciousness is often observed in dementia praecox' (Diefendorf, 1918, p. 58).] Bleuler was equally familiar with work on dissociation in hysteria by Pierre Janet (1859–1947). However, Bleuler just as often appears to have formed his impression from clinical observations. He wrote, for example: 'Naturally such patients must speak of themselves in one of their two versions or they may speak in the third person of the other two. This sort of reference ... is the expression of a real alteration in personality. But even when such a splitting cannot be demonstrated ...' (Bleuler, 1911/1950, p. 144). All this is not entirely surprising. Liberally used throughout the description of symptoms in schizophrenia is his still evolving notion of the complex (a group of ideas that are strongly affectively charged and influential towards other psychic processes). In 1906 his understanding of the unconscious complex was expressed as follows:

> There is ... no difference in principle between unconscious complexes and these several personalities endowed with consciousness. When an unconscious complex associates to itself an increasing number of the elements of the ordinary ego, without linking itself with the ego as a whole, it becomes finally a second personality (1906/1969, p. 291).

In addition to his 1911 comments, Bleuler does express caution. He notes: 'However, they probably are never completely separated from each other since one may communicate with both' (1911/1950, p. 147). Yet despite this caveat, the evidence is clearly overwhelming. There can be no mistaking that the idea of schizophrenia exhibiting more than one personality can, in fact, be rooted to Bleuler's conceptualisation of schizophrenia. And it foreshadows other people's subsequent usage of schizophrenia in this way. Moreover, from 1911 onwards, schizophrenia came to be regarded in the public imagination as a synonym for this theorised accessory symptom. More insidiously, schizophrenia also became linked to the metaphor of the Jekyll and Hyde personality. And, as will be made clear, in the twentieth century, the psychiatric profession and members of allied disciplines played no small part in all of this.

Early Usage

One early usage of both metaphors can be traced to a 1916 interview in the *Washington Post*, with the then president of Clark University, G. Stanley Hall. The interview centres on American concerns over neutrality in the ongoing First World War. The article is titled, 'He Calls it Schizophrenia, and Places Blame on War'. The subtitle of the article reads:

> Dr. G. Stanley Hall, President of Clark University, Finds he Suffers From 'Split Soul' as Result of Trying to Be Neutral—Clouds of Egotism Have Melted—Germans Throw Away Works on Superman and French Return to Christianity (*Washington Post*, 1916, p. A5).

The anonymous journalist in question opens the article rhetorically, asking the reader, 'Have you got schizophrenia? If you have, it is more than likely the world war gave it to you'. Further into the paragraph we find, 'schizophrenia means a split soul or mind, and good Americans are likely to get it in preserving their neutrality toward the warring nations'.

This appears to be the earliest recorded use and definition of the word in the informal sense by someone outside of the mental health profession. It is also the first ever appearance of the word in the *Washington Post*. The journalist is receptive to the new term and is content to act as the conveyor of this neologism. The journalist also clearly separates his/her own thoughts from that of Dr Hall, and quotes verbatim the exact words of his/her questions and those of Dr Hall:

> 'Schizophrenia', Dr. Hall told me, 'is a term much used by psychologists to describe a divided mind, of which the Jekyll–Hyde personality is one type. I was made in Germany, and everything I am I owe to German scholarship, for I spent some of my student days there. At the same time I deplore the militarist spirit. It is in trying to reconcile these conflicting tendencies in myself that I have developed schizophrenia or split soul' (ibid).

First, Hall's comments then, if accurately transcribed, apparently represent the earliest recorded use and definition of schizophrenia in the informal sense by someone *inside* a psychological profession. Second, his passing it to the journalist marks the clear transference of an informal usage of the word schizophrenia from within the profession to outside the profession.

We have then an eminent psychologist, who understood German, defining to a layperson, the meaning of schizophrenia. 'Much used by psychologists', he described it as a split soul, divided mind, and even a Jekyll–Hyde personality. The following question has to be asked: How could a serious professional like Hall deliberately introduce to a journalist an improper definition or caricature of schizophrenia? Could Hall *not* have said, 'I am in two minds about the war', or 'I am indecisive and emotionally torn'? The most plausible answer to such a question is that Hall was not deliberately misusing the term in some new way. Most likely, he was, in fact, using it in a sense that psychologists were using it at the time.

Hall's claim that the term was 'much used by psychologists' suggests that he believed his understanding of the word was familiar to co-workers. Hall's area of expertise was education and developmental psychology—not schizophrenia or dementia praecox. However, he was frequently in contact with work and people related to dementia praecox/schizophrenia. Hall had previously sent his students to study with the dementia praecox researcher Adolf Meyer, at Worcester State Hospital for the Insane, from 1893 to 1902. He had also invited Meyer to the celebrated 1909 Clark University Conference, to which Hall had also famously invited Freud and Jung. In 1910, Hall had also published Jung's lectures on association tests. These were conducted on cases of dementia praecox. And he had instructed his fellows at the Children's Institute to work on the association test. Additionally, in late 1912, Hall had conducted face-to-face meetings over the possibility of establishing a journal with psychiatrist William Alanson White and physician Smith Ely Jelliffe (Ross, 1972). Both men had research interests in dementia praecox. In particular, this was the same year White had published a translation of Bleuler's 'Theory of Schizophrenic Negativism' (Bleuler, 1912).

Hall had ample opportunity then to absorb the latest characterisation of schizophrenia from colleagues. Certainly, psychiatrists are known to have laughed at the name schizophrenia. An undated letter to E.E. Southard from August Hoch states that, 'When I read my review of Bleuler's schizophrenia at the New York Psychiatrical [sic] Society, all of them made a lot of fun of the term' (Southard and Noll, 2007, p. 502). Hoch himself (who in 1912 reviewed Bleuler's work in the *Psychological Bulletin*) thought the term uncouth when he first heard it. However, he noted that it was remarkable what one could get used to (Hoch, 1912; Southard and Noll, 2007).

Finally, as noted, Hall also explicitly stated that the 'Jekyll–Hyde personality' is one type of schizophrenia. As such, in addition to the idea of a 'split soul' Hall introduces the idea of schizophrenia as 'Jekyll and Hyde'. It is not clear if there were any explicit antecedent factors at play in Hall's conceptualisation of schizophrenia as Jekyll–Hyde. There doesn't particularly need to be. He may simply have been translating the concept into something the popular imagination could readily understand. All the same, it may not be insignificant that Hall's comments postdate the publication of *A Mind That Found Itself* by Clifford Beers in 1908. Among other things, this ground-breaking asylum memoir described a number of institutional abuses by an assistant physician that Beers nicknamed 'Jekyll Hyde' (Beers, 1908). If Hall was directly or indirectly aware of the damming contents of these famed memoirs, as seems plausible, his reading of schizophrenia reverses this damming characterisation entirely. It was now the person suffering mental illness who was Jekyll–Hyde. Whatever the case, Hall certainly did not reject the characterisation of schizophrenia as Jekyll–Hyde or a divided mind—quite the opposite. He passed such ideas to the public. And, as we shall see, for much of the twentieth century such ideas would remain stable in the public conception of schizophrenia.

In the early twentieth century schizophrenia was often taken as a synonym for dementia praecox. Hence, it is not surprising to find that in 1919 the *Washington Post* also ran an article along similar lines. The article, titled 'Was Jekyll and Hyde', told the story of a successful university student shot dead while leading a double life as a burglar. A 'noted', but anonymous, psychologist speculates that the student, 'was probably a victim of Dementia Praecox' (*Washington Post*, 1919, p. 4). To what extent the psychologist was aware of the Jekyll and Hyde theme to the journalist's pending story is unclear. But once again the usage of Jekyll and Hyde is associated with the involvement of a psychologist.

The transfer of schizophrenia as variants of a split personality continued unabated from professional sources to the public domain. And over time this transfer became increasingly less ambiguous. This can be seen five years later in a *Washington Post* article entitled, 'How to Keep Well; Schizophrenia'. This time the definition is passed to the reader, from *Stedman's Medical Dictionary*, via a respected intermediate authority figure, Dr Evans, a weekly health columnist:

> During the next several months the country is liable to hear much of schizophrenia. The definition of this work [sic], given by Stedman, is a condition marked by splitting of the personality, or

> intrapsychic ataxia. A second definition of it is dementia praecox (Evans, 1924, p. 10).

Significantly, it is noteworthy that 'splitting of the personality' has moved within the medical profession. It has moved from being considered as an uncommon symptom of schizophrenia to being part of a publicly endorsed definition of schizophrenia. Interestingly, the article goes on to debate the possible existence of a schizoid personality in several past presidents. It quotes the psychiatrist Abraham Arden Brill, addressing the APA, who notes:

> the marked tendency of the schizoid to a splitting of personality between reality and fancy often leads to an insane condition (ibid).

Brill's reported comments—a merging of Bleuler's separate theorisation of splitting of personality and thoughts on autism—would have done little to curb the informal conception of schizophrenia (although, admittedly, Brill seems mostly concerned with stoking his fellow professional's ire by retrospectively diagnosing several past American presidents, including Lincoln, as schizoid). Brill's comments would have had additional influence because he was among the first practising American psychoanalysts. And he was highly influential in the establishment of New York city as the psychoanalytical centre of the USA (Richards, 1999). Moreover, he was an assistant to Bleuler for several months in 1907. He was also the translator of Bleuler's 1924 *Textbook of Psychiatry* (Bleuler, 1916/1924).

In the 1924 translation of Bleuler's *Textbook of Psychiatry* a discussion of personality transformations occurs in the section on accessory symptoms. Bleuler again refers to splitting:

> His recollections are split into two or more parts; the one set of his experiences he ascribes to the real John Smith, the other to his new personality which was born in Charenton and is named Midhat Pasha. Others become a new personality at a definite moment (1916/1924, p. 393).

In a discussion on more general psychopathology of personality, Bleuler also discussed schizophrenia. He notes: 'His personality can divide itself; now he acts and thinks like a great man, now like a scholar ...' (1916/1924, p. 141). Similarly, he noted that 'schizophrenia produces different personalities existing side by side' (1916/1924,

p. 138). Again, as in his 1911 text, Bleuler's observations are qualified. He talked of other types of personality disturbances such as transformations of personality. Furthermore, in respect to multiple personality he stated, 'As a matter of fact, cases of pure dual personalities are very rare' (ibid). Nevertheless, the idea of the 'split personality' remained within his work.

By 1931 Brill was again involved in a debate concerning schizoid presidents. He is noted by the *Washington Post* as stating that Lincoln 'was a dual personality who rigidly controlled his baser nature', and that 'Lincoln had a schizoid or Dr. Jekyll and Mr. Hyde personality' (1931, p. 2). Self-evidently, such a professional debate, 'at a joint session of the American Psychiatrist [sic] and Psychoanalytic Associations' (ibid), concerning one of America's more renowned presidents had not escaped public attention. The analysis was declared insulting by a Dr J.L. Moreno, who objected to Brill's use of retrospective diagnosis. But Moreno appears to have made no effort to refute the possibility of the existence of a schizoid, Jekyll and Hyde, dual personality per se.

In 1924, the same year that Bleuler's *Textbook of Psychiatry* appeared, the linkage between split personality and schizophrenia surfaced in a Chicago murder trial. The strangled victim, Bobby Franks, was murdered by two academically gifted young men from wealthy and socially established families. Their names were Richard Loeb and Nathan Leopold. Their apparent motive was to commit the perfect crime. This revelation shocked the public, who struggled to comprehend the mental processes of the two men. The crime became an international sensation. A large number of psychiatrists testified at the trial. One publisher even offered Freud half a million dollars to examine the accused (Geis and Bienen, 1988). Various psychological explanations were proposed, including the diagnosis of prodromal schizophrenia/dementia praecox, and many psychiatric theories and observations received widespread coverage and analysis. As such, during the trial defence lawyer Charles Darrow (later of Stokes trial fame) gently mocked the psychiatric profession's latest concept. And this was duly reported: '"*Schizipathic*" he called it. "*Schizileptic*" and "*schizzy-what-ever you call it*"' (*Washington Post*, 1924a, p. 3). [Earlier, in a 1907 murder trial, a defence lawyer, Delphin Delmas, had similarly mocked dementia praecox with 'dementia Americana' (Noll, 2011).]

During the trial, any references to schizophrenia as a split personality would have been chewed over by an insatiable public. And they would have reinforced the informal conception of schizophrenia in the public imagination. And indeed there were references to the split personality.

For example, the trial involved William A. White, president of the APA (1924–25). As noted earlier, White was a translator of Bleuler's work: *The Theory of Schizophrenic Negativism* (1912). The prosecution asked White about Loeb's 'split personality' and tried to get the doctor to say whether he was afflicted with dementia praecox (*Washington Post*, 1924b, p. 1). Another psychiatrist, Dr Harold Singer, was reported as having admitted 'that split personalities may develop a psychosis—go crazy—that such psychoses are developed most frequently in the adolescent period—and that such insane persons—schizophrenics is the term the doctor used—frequently commit crimes, even murder, without a motive' (*Washington Post*, 1924a, p. 3). A third psychiatrist, Dr Healy, is quoted as stating, 'To my mind the crime itself is the direct result of diseased motivation of Loeb's mental life. The planning and commission was only possible because he was abnormal mentally, with a pathological split personality' (Higdon, 1999, p. 217). As such, we see here immense public exposure to the psychiatric profession's prevailing conceptualisation of schizophrenia and dementia praecox. Crucially, there appears to be no records of psychiatrists in the trial disavowing the 'split personality' as terminology. This is not surprising. Four years later, Carl Jung, whose 1902 dissertation had dealt with a medium exhibiting a second personality (Decker, 1986), was still writing:

> in schizophrenia the normal subject has split into a plurality of subjects, or into a plurality of autonomous complexes. ... The simplest form of schizophrenia, of the splitting of the personality, is paranoia. ... It consists in a simple doubling of the personality, which in milder cases is still held together by the identity of two egos ... (1928/1972, p. 226).

Given such assertions were not unusual, it seems understandable as to why a 1931 Kansas newspaper discussing 'Colorado's macabre scarecrow man' could sensationalise much about a man found standing cruciform in a corn field but seemingly felt little need to embellish the actual diagnosis: 'Psychological experts, proclaim him a victim of schizophrenia or split personality' (*Ogden Standard Examiner*, 1931, p. 31). By now, we can also find a sharp decline in the usage of the diagnosis of multiple personality itself in favour of schizophrenia (Rosenbaum cited in Putnam (1989)). [Rosenbaum dates the transition to around 1927. Criticism by William McDougall appears to have played a role in the decline of multiple personality as a diagnosis; some thought it an artefact of hypnosis (Putnam, 1986, 1989).]

In passing, it is worth noting that the first appearance of the informal usage of schizophrenia in the English language has been attributed by the *Oxford English Dictionary* to T.S. Eliot's *Use of Poetry and Use of Criticism*. There Eliot can be found stating:

> For a poet to be also a philosopher he would have to be virtually two men; I cannot think of any example of this thorough schizophrenia, nor can I see anything to be gained by it (1933, p. 99).

But it is clear from the preceding discussion that the informal metaphor had long since been established and would continue to be perpetuated. Similarly, usage of Jekyll–Hyde seems to have continued unabated as well. The *New York Times*, for example, would run an associated press article from London in 1941. This perpetuated the metaphor in discussing the perceived insanity of the Nazi deputy Rudolf Hess. The German had, seemingly unbeknownst to an outraged Hitler, parachuted into wartime England to conduct peace negotiations.

> LONDON, May 18 (UP)—The Marquess of Donegal, writing in The Sunday Dispatch under the headline 'Complete Explanation of Hess', said that the Nazi deputy leader was suffering from schizophrenia or a Jekyll–Hyde personality (*New York Times*, 1941, p. 4).

Academic reports were little different. In *The Biology of Schizophrenia*, Harvard's R.G. Hoskins could also claim, 'Rather literally several warring persons exist in the same body and the patient is truly bewildered as to which one to accept as "I"' (1946, p. 92). In 1950, New York psychotherapist Margaret Naumburg reported, in *Schizophrenic Art: Its Meaning in Psychotherapy*, that she had postponed proposed electric shock treatment after her pressurised patient Harriet produced therapeutic data: an image titled 'Showing How the Split in the Girl's Personality is Being Healed' (1950, p. 134). In the same year, Davis could also write up a case of "schizosis" with dual personality (Davis, 1950).

The striking metaphors of schizophrenia as a split/Jekyll and Hyde personality came into being in this way then and was evidently not easy to displace. Indeed, the belated translation of Bleuler's core text in 1950 would serve not to introduce the idea of the split personality but, in fact, to reinforce it (in this sense, the transformation of a historical document can serve to perpetuate its own mythology). In the foreword to the translation, Nolan D.C. Lewis further strengthens the idea,[3] adding that Bleuler,

> considered the fundamental symptoms to represent a splitting of the personality which he designated 'schizophrenia' (Bleuler, 1911/1950, p. iii).

But of course Bleuler hadn't been *that emphatic*. Indeed, a few pages later the reader can find Bleuler emphasising the splitting of psychic functions not the splitting of personality.

Decline of the Metaphor

The demise of the informal metaphors split personality and Jekyll and Hyde within the profession appears to have been gradual, and the reasons are probably multiple. The most notable intellectualisation of the change can perhaps be tracked to the writing of James May in 1931. As APA president, May had previously chaired a 1928 committee, whose members had facilitated the compilation of *The Standard Classified Nomenclature of Disease*. This was a general classification of diseases that included psychiatric classification. In doing so, the manual upheld an ongoing tradition in psychiatric classification already evident in 1918 (this will be visited in detail later). It treated Emil Kraepelin's dementia praecox and Bleuler's schizophrenia as synonyms. Writing just after the publication of the manual, May formally stated that he had viewed the two concepts as similar. Indeed, May's committee had enforced synonymy. For May, there was only one essential difference:

> It is of some interest that schizophrenia, so-called, still includes the simple hebephrenic, catatonic, and paranoid forms originally described by Kraepelin. The one essential point of difference insisted upon is the splitting of personality as the characteristic feature of the disease (1931, p. 438).

As such, May, in what can be read as a post-hoc justification of the work of the committee, and perhaps earlier committees, later proceeded to downplay the role of the splitting of personality. Instead, and as might be found in descriptions of dementia praecox, he merged it into a disorganisation of intellectual mechanisms:

> the dissociation or splitting, when it occurs, is primarily due to a disorganisation of the purely intellectual mechanisms which results in a corresponding incoordination of their functions (1931, p. 440).

Such a reading meant that an important leader in official North American psychiatry no longer rationalised schizophrenia as a process of intellectual disorganisation *accompanied by* a split personality. Instead, intellectual disorganisation now subsumed the split personality—a characteristic feature of the disease (as May put it). This merger seemingly helped May to bridge dementia praecox and schizophrenia conceptually, and treat them as more or less equivalent. As APA president, May's pronouncement would have carried some weight. It was also—subject to interpretation—partially in keeping with some of Bleuler's 1911 writing and, of course, that of Kraepelin. Furthermore, any side-lining of the split personality caused no overt problems in classification per se because there were still other options for curious cases. Psychiatric manuals, for example, had always additionally allowed dissociation phenomena to be placed under categories such as conversion hysteria. Simultaneously, and working in the opposite direction, the latter possibility may have further prompted a reconfiguration of schizophrenia. In 1940, for example, Osborne would observe that Bleuler's emphasis on 'a personality split as the prime symptom' (1940, p. 1078) was problematic because dissociation was seen in other disorders.

By 1955, another significant change can be found, this time in Curran and Partridge's introduction to *Psychological Medicine*. The British psychiatrists noted that 'Schizophrenia does not give rise to the Jekyll and Hyde personalities of the popular press; these are usually psychopaths in whom a surprising combination of good and bad qualities or behaviour is shown' (Curran and Partridge, 1955, p. 193). In doing so, the authors had attempted to clarify the definition of schizophrenia. They had done so while seemingly equating the Jekyll and Hyde personalities 'of the popular press' with the classification of psychopath. They further cautioned: 'In schizophrenia a better analogy is a widespread splintering of the mind, rather than a relatively simple split of the alleged dual personality type' (1955, p. 193).

These comments are particularly noteworthy because they are absent from earlier editions of the same book, *Psychological Medicine*, published in 1944 and 1946. Both of these had read, 'The main features are progressive introversion, splitting of personality, and paranoid symptoms' (Curran and Guttman, 1944, p. 36; 1946, p. 47). In citing a need to move away from the 'Jekyll and Hyde personalities of the popular press', as opposed to those of the psychiatric profession, Curran and Partridge can be seen as early contributors to a myth. This was a myth that would come to declare the 'schizophrenia/Jekyll and Hyde personality' as a public misconception. What exactly prompted the 1955 revision is unclear. But it is possible that the book's more nuanced understanding of schizophrenia follows the 1950 translation of Bleuler's work into English.

The tendency by author's such as Curran and Partridge to transition away from conceiving schizophrenia as split personality was still not universal. In fact, the informal professional usage continued and remained international. For instance, in 1961, V.A. Gilyarovsky (chair of the USSR Society of Neuropathologists and Psychiatrists) observed the prefix *schizo* denotes 'the same "splitting" of the personality, which is undoubtedly a characteristic feature of the disease' (1961, p. 286). In this case only the use of quote marks around splitting suggests an implicit qualification of the term. Elsewhere, R.D. Laing (1960/1990) would speak of the split in the schizoid individual's being. While Bleuler's own son, Manfred, also a psychiatrist, would continue to perpetuate and reinforce the loose usage as late as 1966: 'The result is the splitting of the personality, of the whole inner life, which we encounter in the schizophrenic. Schizophrenia is to be understood as a faulty development of the personality ...' (Bleuler, 1966, p. 3).

Nevertheless, psychiatry was slowly moving away from the idea of the split personality as representative of schizophrenia. It is possible that the demise of the 'split personality' was further due to a perceived need to distinguish schizophrenia from multiple personality. After its ostensible elision into schizophrenia around 1927, multiple personality had re-erupted once again into cultural consciousness in the second half of the twentieth century. This occurred with the 1957 publication of *The Three Faces of Eve* (Thigpen and Cleckley, 1957) and the eponymous film starring Joanne Woodward, both of which followed an earlier publication by Thigpen and Cleckley (1954) on multiple personality, in the *Journal of Abnormal and Social Psychology*. Hence, we can indeed find Hilgard's 1962 *Introduction to Psychology* cautioning that in schizophrenia 'the split is *not* usually into multiple personalities, as in the Eve White Case' (1962, p. 525). Despite the emphatic 'not', the use of the adverb 'usually' nevertheless still admits the possibility that sometimes the split *was* into multiple personalities.

It may further have been necessary to distinguish it from any possible misinterpretation of results of early split-brain experiments by Gazzaniga et al. (1962). [Although the possibility that two separate minds might exist in schizophrenia, owing to corpus callosum abnormalities, was still voiced by Nasrallah as late as 1979 (Greenberg, 1979).] Perhaps equally problematic were negative portrayals of schizophrenia in works like Truman Capote's *In Cold Blood*, a much-lauded 'true account' of the slaying of a family in Orange County. Capote noted of one 'schizophrenic' killer Andrews: 'inside the quiet young scholar there existed a second, unsuspected personality, one with stunted emotions and a distorted mind through which cold thoughts flowed in cruel directions' (Capote, 1966/2005, p. 304). Such portrayals may have sat uneasily and inconsistently alongside a growing cultural drift to seeing the patient as a victim of culture.

Clearly, though, post-1950 the tide had turned. By 1968, Silvano Arieti, probably one of the most influential thinkers in mid-twentieth-century schizophrenia research, was cautioning in the *Encyclopaedia Britannica*: 'It was Bleuler who coined the term schizophrenia. *By this term he did not mean a split, divided, or double, personality*, as is popularly believed but a lack of coordination between various psychological functions' (1968, p. 1162, my emphasis). Arieti did not say why it was 'popularly believed'.

Later, in a similar dislocation of the origins of the myth, Harvard's Patrick O'Brien, in *The Disordered Mind*, explained that, 'Eve in *The Three faces of Eve* and Dr. Jekyll in *Dr. Jekyll and Mr. Hyde* were not schizophrenic' (1978, p. 50). Furthermore, he stated that 'if clinicians are somewhat vague and confused in what they call schizophrenia, the general public is even vaguer and more confused and prey to an enormous number of superstitions and misunderstandings' (ibid). Instead, O'Brien revealed to his readership that the fictional character 'Dr. Jekyll was suffering from hysteria, not schizophrenia' (1978, p. 10). Ironically, in 1911 Bleuler had acknowledged that hysteria itself was partially merged into schizophrenia. And as historian Marc Micale has pointed out in *Approaching Hysteria*, Bleuler's schizophrenia, incorporated 'within itself several components of the old hysteria, now signified under new names and camouflaged in different theoretical surroundings' (1995, p. 173).

By 1980 Bleuler's concept of schizophrenia had long been rejected in favour of other models of schizophrenia. For example, through Kurt Schneider's supposedly pathognomic first-rank symptoms (see later) the psychiatric profession had moved to delineate readily identifiable criteria, none of which mentioned 'split personality'. Indeed, as a reflective Jean Garrabé (2003) would later muse, in the DSM-III the dissociation or spaltung that Bleuler deemed *central* to his conception of schizophrenia had been entirely extracted from the concept of schizophrenia. It was now in another category: dissociative disorders. Whatever schizophrenia had been, it was no longer Bleuler's schizophrenia qua splitting.

Fiction

Those diagnosed with schizophrenia had become entangled in the fictional Jekyll and Hyde. Yet, in a final twist, fiction could equally find itself entangled with what Jelliffe (1927) would call the nosological fiction that is schizophrenia. For example, Moss and Hunt, in their 1932 work *Foundations of Abnormal Psychology* diagnosed Shakespeare's Ophelia as suffering from hebephrenic dementia praecox (Moss and Hunt, 1932; see also Wilson, 1951). While, Redlich could, in a slightly different way, declare that,

> Snow White fell into 'catatonic stupor' because she ate a poisonous apple given to her by her 'schizophrenogenic' stepmother (1952, p. 22).

Claude and Levy-Valensi would attempt to find schizophrenia in Balzac's *Comédie Humaine* (Lewis, 1936). F.G. Crookshank (1930) would cast Don Quixote as a 'flat schizoid'. And Paul C. Squires, in the *Psychoanalytic Review*, would see Fyodor Dostoevsky's fictional murderer Raskolnikov as schizophrenic. He argued that the name Raskolnikov was in itself of the greatest psychological significance. For Squires (1937) it meant a dissenter, split off, in a state of schism, schizophrenic. Similarly, Aldo Calanca (1974) would query Sartre's Antoine Roquentin (Nausea) as a case of simple schizophrenia.

Other disciplines would do the same as twentieth-century society universalised its latest concept of madness. And in this sense texts continue to be 'written' long after their authors' demise. Hence, Herman Melville's 'self-divided' Bartleby (the Scrivner) could be considered variously schizophrenic with catatonic features or schizoid by Beja in *The Massachusetts Review*. This occurred even though Beja was aware of many critics' distaste towards treating artefacts of the imagination as people (Beja, 1978). By 1985, Ferrier continued to marshal schizophrenia evidence from the behaviour of Shakespeare's 'Poor Mad Tom'. He did so, while admitting the suggestion that Tom as a case of chronic schizophrenia was not only contentious but untestable (Ferrier, 1985). In its extreme form, we can even find the term schizophrenia reflecting back from the mirror of literature and being applied to whole groups of authors. For instance, the work of young twentieth-century authors was described by *Revue des Lectures* as being characteristic of schizophrenia (Anon, 1932).

Ostensibly such apodictic and nonchalant didactic exposition shed light on both society and madness (and increased the authority and prestige of those who entertained it). Yet, in hindsight, unfalsifiable and often anachronistic diagnostic projection into fiction—including the ever-versatile fairy tale—ran in conjunction with ongoing epistemic difficulties experienced by those who sought to conceptualise schizophrenia. And at least in one sense, canonical fiction served as a scaffold for an unstable concept. Such a claim will become clearer as we further explore through definition and classification just how unstable the concept of schizophrenia was in the twentieth century.

3
Definitions of Schizophrenia

One way schizophrenia researchers attempted to pin down their conception of twentieth-century madness was through definition. So we now turn to examine definitions of schizophrenia. But before proceeding, it is worth remembering that the contemporary reader, and those involved in schizophrenia research, bring to any such history an entirely different mindset regarding definitions per se in comparison with early twentieth-century psychiatrists. For that generation the linguistic turn of philosophy was never really apparent. And controversy over the nature of definition itself would not emerge as a major theme in psychiatric research until approximately the second half of that century.

Instead, early schizophrenia researchers by and large saw the process of definition as a valid method of discussing a concept. And rather than analysing this form of representation itself, they limited themselves principally to debates and to concerns over the content of definition. That psychiatrists did not recognise the limitations of definition is unsurprising. It was only in 1950 that a major analysis of this form of representation was carried out by Richard Robinson in *Definition* (1950). In that work, which used as examples 18 definitions of 'definition', the author identifies several species of definition. In doing so, he revealed that it has been variously argued that a definition need not necessarily be brief. An intellectual or scientific endeavour might equally work towards a definition, as opposed to commencing with one. And many definitions of the same term might be better than one alone.

Furthermore, Robinson observed that it has been maintained that some things are indefinable. That it has been suggested that the whole procedure is worthless or vice versa. And that a definition cannot be either true or false because it is not a statement but rather a command;

not a proposition but a proposal. Finally, in this list of charted controversy, he reminded readers of the sticky question, as to whether definitions apply to things, words, or concepts (ibid).

No definition then will capture the variety of meaning possessed by schizophrenia at a given time. Nevertheless, we can examine what is included in a given definition and collate these with other definitions in a chronological manner. In doing so, we improve our theoretical understanding of changes in schizophrenia conceptualisation. We also improve our understanding of its complex ontological nature and of the ontogenesis of what would become known as operational definitions. In doing so we chart what Foucault would perhaps call the 'ceremonial space' of definition. Such an investigation helps lay foundations for a broader comprehension of the representation of psychiatric concepts. It also helps us understand how they are maintained and the transhistorical manner in which they are subject to change.

Bleuler

First, let us retrace our steps a little. As we have seen, Swiss psychiatrist Eugen Bleuler, of the Burghölzli clinic in Zürich, first introduced the term schizophrenia in 1908. And in that brief paper, he presented the following working definition or proposal. It comes as close as he ever came to explaining the derivation of *schiz* (to split) and *phrene* (mind):

> I believe that the tearing up or the splitting of the psychic functions is the outstanding symptom of this whole group [*Ich glaube namliche das die Zerreißung oder Spaltung der psychischen Funktionen ein hervorragendes Symptom der ganzen Gruppe sei*] (1908, p. 436, my translation).

In the same paper, Bleuler went on to present information on the recovery rates of his patients. He believed these to be much superior to those previously estimated for dementia praecox. Instead of deteriorating, more patients than hitherto thought now showed indications of recovery. Bleuler's proposal was, in the author's own mind, not so much at this stage a definition of schizophrenia. It was more a *redefinition* or reformulation of dementia praecox, which Kraepelin had characterised before 1908 as,

> The complete loss of mental activity, and of interest in particular, and the failure of every impulse to energy, are such characteristic

> and fundamental indications that they give a very definite stamp to the condition. ... Together with the weakness of judgment, they are invariable and permanent fundamental features of dementia praecox, accompanying the whole evolution of the disease (1904/2002, p. 26).

In 1909, A.A Brill would introduce the term in the *American Journal of Insanity*, again casting schizophrenia as a synonym to dementia praecox. He did this with little elaboration. But he did thank Bleuler for his optimism. He also thanked him for repudiating the meaningless term dementia praecox, which was neither 'a dementia nor a praecox' (Brill, 1909, p. 69). [Around the same time, the Meyerian school was also transitioning from Kraepelin's dementia praecox to more psychogenic ideas (Noll, 2011).] Elsewhere, in the *American Journal of Psychiatry*, E.E. Southard (1910) would, in passing, briefly parrot Bleuler's definition: tearing up or splitting of the psychic functions (*Zerreißung oder Spaltung der psychischen Funktionen*). And a year later, Stedman's *A Practical Medical Dictionary* also equated it with dementia praecox: 'schizo, I cause to curdle, + phren mind. Dementia praecox' (1911, p. 777). At this stage, there was no substantive discussion of the merits of the term in the Anglophone literature (which this investigation is largely restricted to). And it is not until much later within the Anglophone literature that definition, or indeed the very idea of producing one, received critical scrutiny. [Note: I will not deal here with ostensive definitions, save to note that even Bleuler could dubiously state, 'Sometimes one can make the diagnosis almost with certainty from listening to a short piano recital' (1911/1952, p. 88). Elsewhere, Bleuler states he could make a diagnosis in many simply from their 'will-o'-the-wisp-like gait' (1911/1952, p. 171).]

In 1911, Bleuler attempted his first stipulative definition of schizophrenia. It occurred in his classic text *Dementia Praecox or the Group of Schizophrenias*, in the subsection entitled 'Definition of the Disease':

> By the term 'dementia praecox' or 'schizophrenia' we designate a group of psychoses whose course is at times chronic, at times marked by intermittent attacks, and which can stop or retrograde at any state, but does not permit a full *restitution ad integrum*. The disease is characterised by a specific type of alteration of thinking, feeling, and relation to the external world which appears nowhere else in this particular fashion (1911/1952, p. 9, original emphasis).

The first thing to note here is that although more optimistic than Kraepelin, Bleuler did not consider that a patient could make a full recovery. Second, the theorised 'group of schizophrenias' was already contracted to schizophrenia for ease of usage. The subtlety of this last point in particular was lost on many psychiatrists. Many would simply see schizophrenia as a singular disease, rather than a group of diseases (Curran and Guttman, 1946). Bleuler also now demoted the idea of the 'tearing up' (*Zerreißung*) of the psychic functions, first expressed in 1908. And he stated that he called dementia praecox schizophrenia because the,

> 'splitting' of the different psychic functions is one of its most important characteristics (Bleuler, 1911/1952, p. 8).

From this point on his emphasis was not on tearing *and* splitting, but on splitting alone. Furthermore, *within the definition*, he noted, 'In every case we are confronted with a more or less clear-cut splitting of the psychic functions. The personality loses its unity; at times different psychic complexes seem to represent the personality' (Bleuler, 1911/1952, p. 9). And, as we have seen already, this point of definition was later reinforced with comments on the personality such as:

> Naturally such patients must speak of themselves in one of their two versions or they may speak in the third person of the other two. This sort of reference is here not merely an unusual or awkward figure of speech ... but is the expression of a real alteration in personality. But even when such a splitting cannot be demonstrated ... (1911/1952, p. 144)

We see then how a clear line of interpretation opens up, whereby the possibility of understanding schizophrenia moves from a tearing up, to splitting, to splitting of the psychic functions, to splitting of the personality. As we shall see shortly Isador Coriat appears to partially follow this line of reasoning as early as 1914 (Coriat, 1914). And, of course, for some such as G.S. Hall, this line of interpretation even went as far as reaching Jekyll and Hyde. Many definitions follow a somewhat similar but different line of interpretation. Some, for example, appear to conceptualise the splitting of psychic functions as the splitting of mind (Warren, 1934), or even the heart as the seat of thought (Baynes, 1940, p. 21). Bleuler's assistant, Jörger (1918), could even speak of 'split being' (*Gewaltsein*) in an architect.[1]

A Broad Concept

In addition to an emphasis on splitting, it now became evident for the first time that what Bleuler proposed was much broader than dementia praecox. In 1912 Piquemal would describe schizophrenia as infinitely vast, which perhaps overstates things (Piquemal, 1912). But quite how broad the concept was, or indeed became, is open to interpretation. In his 1912 review of Bleuler's work for the *Psychological Bulletin*, August Hoch, noted:

> Bleuler comprises in his book a great many cases which others would not include in the group or groups of dementia praecox, so that his analysis refers in reality a great deal to the symptomatology of the functional psychoses [sic] (1912, p. 169).

In 1926, admitting the concept of schizophrenia lacked precision, Bleuler's own translator, Henri Ey, would also acknowledge surprise that he could see pre-existing concepts in latent schizophrenia: the majority of the paranoias, incurable hypochondriacs, a lot of cases of moral insanity, and amentia or confusion mentale (Ey, 1996). Similarly, in 1926, during a conference attended by Bleuler, Bordeaux's Anglade would argue that Bleuler had drawn on the psychiatric map, a schizophrenic confederation, whose borders were uncertain (Bleuler and Claude, 1926/2001). Understandably, then, individual readings could by idiosyncratic. Isador Coriat, for example, would take latent forms to include individuals with oddities of character. These included reticence, seclusiveness, and other abnormalities of mental make-up (Coriat, 1917). But, for the most part, interpretations simply varied. Elsewhere, Harry Stack Sullivan, for example, drew attention to the fact that schizophrenia was broad enough to include 'the clinical entity hysteria' (1931, p. 525), which perhaps fits with earlier comments by Bleuler that a large number of women whom he considered schizophrenics passed for hysterically insane in other places. Michel Foucault thought the essential difference was that Bleuler had extended dementia praecox to include certain forms of paranoia (Foucault, 1954/1987). This fits a declaration by Bleuler that the term paranoid was more expansive than Kraepelin's earlier conceptualisation of 'dementia paranoides'. And it fits Bleuler's declaration that, 'I am unable to narrow down my concept of paranoia until it corresponds with Kraepelin's' (1911/1952, p. 280). Yet other researchers have thought that the distinguishing difference between dementia praecox and schizophrenia was dissociation (May, 1931;

Warren, 1934). Some welcomed the extension (Skottowe, 1940). Silvano Arieti (1955), for instance, would approvingly detail the broader forms. These were not just latent schizophrenia but 'Psychoses which arise in psychopathic personalities, alcoholic hallucinoses [sic], prison psychoses and cases of symptomatic manic-depressive psychoses' (Arieti, 1955, p. 14). However, years later, Joseph Zubin would complain:

> In no other field of medicine has anyone ever solved a problem by extending it. We took the old dementia praecox concept and broadened it through Bleuler's efforts, and later through his followers', to the kind of spectrum that makes it impossible to try to do anything specific with it as a disease entity. I can't understand why that happened ... (1961, p. 202).

Bleuler was reproached about the concept's broadness. However, he reportedly stated that it was a fact that there were more horses than elephants, more colds than typhoid fevers (Southard, 1914). In any case, Bleuler was not alone in being accused of expansion. In 1910, Adolf Meyer had similarly grumbled of expansion by Kraepelin (1910, p. 276). While Meyer's own conception was later described as 'anti-nosology' and as an abusive extension of the concept—along with Sullivan's concept—and, indeed, American schizophrenia in general (Ey et al., 1977). Nonetheless, when deciding between dementia praecox and schizophrenia the literature as typified by Harvard's F L. Wells in *The Journal of Abnormal and Social Psychology* generally concedes to Kraepelin the tighter description:

> Schizophrenia is only by courtesy a diagnostic entity since it denotes a symptom-complex which, comparably to aphasia or even fever, is observable over an indefinite nosological area. The Kraepelinian nomenclature had from this standpoint a sounder descriptive basis (1946, p. 199).

Recognising Bleuler's enlargement, Kraepelin himself would comment that it remains to be seen whether the term would gain widespread acceptance. And he continued to use the term dementia praecox (Kraepelin, 1913/1919). By 1914, Isador Coriat, in the *American Journal of Psychiatry*, picked up on the extension of dementia praecox. He noted that Bleuler's term was 'much broader than in the usual Kraepelin sense' (Coriat, 1914, p. 679). Despite Coriat's indisputably true observation, the author nevertheless unaccountably defers to Bleuler's claim of conceptual synonymy with dementia praecox:

> Dementia praecox or schizophrenia is defined as a group of chronic psychoses, with outbursts or remissions, each case showing a more or less clear splitting of the personality and disturbances of associations, but without primary disturbances of perception, orientation or memory (1914, p. 679).

He would not be the last to do so. By now, a greater emphasis on Freudian thinking was also emerging.

A Freudian Reading

By 1914, Freud had noted of schizophrenics:

> Patients of this kind ... display two fundamental characteristics, megalomania and diversion of interest from the external world—from people and things. In consequence of the latter change, they become inaccessible to the influence of psychoanalysis and cannot be cured by our efforts (1914/1981, p. 74).

For all that, Freud's censure did not prevent psychiatrists who were interested in both schizophrenia and psychoanalysis from promoting a Freudian approach to theorisation. In 1915, for example, we find that schizophrenia in *Appleton's Medical Dictionary* is given a Freudian reading. The definition is of particular interest as it was partially edited by Smith Ely Jelliffe, who had co-founded the *Psychoanalytic Review* in 1913. As with Coriat's 1914 interpretation, the definition reads schizophrenia as broader than dementia praecox. The dictionary defines the latter as a chronic psychosis of youth with characteristic and bizarre other signs. But in an effort at distinction, it interprets Bleuler's term as dementia praecox and other psychoses:

> a term used by Bleuler including dementia praecox and other psychoses showing marked libido splitting (Jelliffe and Latimer, 1915, p. 741).

However, although the definition purports to be Bleulerian, it does not stem from Bleuler. He had observed that the theory of the 'failure of investing libido in the object' was not without merit. However, it was, 'as yet insufficiently developed' (Bleuler, 1911/1952, p. 370). Instead, Jelliffe's definition (and Bleuler's passing comment) related to a Freudian-inspired attempt by Abraham (1908/1927) to distinguish dementia praecox from hysteria. A Freudian reading was not entirely contrary to Bleuler's thinking. Bleuler had declared after all that, 'An important aspect ... is nothing less than the application of Freud's ideas to dementia praecox' (1911/1952, p. 1). Nevertheless, the emphasis on libido splitting (a divergence of sexual impulses) took liberty with Bleuler's support for Freudian approaches. Such emphasis promoted and extended Freudian thinking at the clear expense of Bleuler's theorisation and arguments. There were clear advantages in doing so. Twinning psychoanalysis with orthodox psychiatry was one way to spread the concepts of psychoanalysis, and gain Freudian thought respectability among psychiatrists. By emphasising marked libido splitting in schizophrenia, this definition facilitated that agenda. That said, Jelliffe would later realise the limitations of psychoanalysis in a mental hospital (Grob, 1994).

Kraepelin's Viewpoint

Moves towards a Freudian conceptualisation of dementia praecox or schizophrenia were all the same nascent. And from as early as 1911, psychoanalysis and indeed psychogenic theories encountered resistance and competition from other theorists (Noll, 2011). Notably in North America, a 1919 translation of Emil Kraepelin's 1913 thoughts on dementia praecox now appeared. In it, Kraepelin expressed regret over the confusion created by the misleading term dementia praecox. Kraepelin nevertheless now further separated the paraphrenias from dementia praecox and offered to psychiatrists a predominantly biological reading in which,

> Dementia praecox consists of a series of states the common characteristic of which is a peculiar destruction of the internal connections of the psychic personality. The effects of this injury predominate in the emotional and volitional spheres of mental life (1913/1919, p. 3).

Kraepelin did not enthusiastically endorse Bleuler's label or conceptualisation (an impossibility, as in the same edition he rejected Freudian theory as 'castles in the air'). However, he took on board aspects of Bleuler's thinking, noting that it remained to be seen if Bleuler's term, or other competing ideas, would be adopted. As such, Kraepelin retained his own concept. And he found it quite natural to speak of 'those schizophrenic disorders which we meet with in dementia praecox' (Kraepelin, 1913/1919, p. 76).

This seems to suggest that Kraepelin, like others, acknowledged the possible existence of schizophrenic disorders both inside and outside the rubric of dementia praecox. Indeed, later in 1920, Kraepelin would note that, 'schizophrenic manifestations can doubtless occur without any brain damage' (1920/1992, p. 523). This was apparently not the case with dementia praecox. He also noted the possibility that other curable diseases 'may sometimes assume the guise of schizophrenic illnesses' (ibid). In any case, by 1913 he now considered dementia praecox to exist in eight clinical groups as opposed to Bleuler who conjectured four (excluding the latent schizophrenic). As such, Kraepelin gave little succour to either Freud or Bleuler.

In places like Argentina, the use of dementia praecox as a diagnosis could still be found as late as 1934 (Eraso, 2010). Nevertheless, Kraepelin's formulations had continued to dissatisfy many (Noll, 2011). As Meyer Solomon, in *The Journal of Abnormal Psychology*, noted:

> No one, in the present state of our knowledge of this problem, can give a simple, short, clearly understood and generally accepted definition of what is meant by the term dementia precox. Why? For the simple reason that the problem has not been definitely solved ... the term is retained for the present to refer to a heterogeneous group of syndromes of many sorts, and in the meantime efforts are being made to understand the conditions now being denominated 'dementia precox'. What will eventually happen, let us hope soon? There will be a dismemberment of the so called dementia precox group ... (1917, p. 197).

Terms employed should have a definite meaning, thought Solomon. They should stand for something specific, something that could be quickly and clearly defined. When terms could not be defined they were not clearly understood.

Schizophrenia Condensed and Reinterpreted

Bleuler, in his 1916 *Textbook of Psychiatry*, largely ignored Kraepelin's latest formulation, and continued to maintain his 1911 perspective. In a highly condensed sketch of the concept, he retained the claim of conceptual synonymy with dementia praecox, and now declared of schizophrenia:

> It is characterized by a specific kind of alteration of thinking and feeling, and of the relations with the outer world that occur nowhere else (Bleuler, 1916/1924, p. 373).

Such a loose characterisation extended and foreshadowed a confusion, which was now growing, as to what precisely usage of the term schizophrenia implied.

In that same year, however, *Lang's German–English Dictionary of Medical Terms* would once more lend support to Bleuler defining *Schizophrenie* as 'f. dementia praecox, schizophrenia' (Meyers, 1924, p. 477). This followed earlier official support for the claim to synonymy by the classification board of the American Medico-Psychological Association. So in the psychiatric domain schizophrenia was already making firm inroads in its claim to being synonymous with dementia praecox. Yet strictly speaking this was inaccurate. A precise understanding of what schizophrenia meant eluded consensus.

Despite official support in North America, Bleuler's reformulation of dementia praecox and his equating of the term with schizophrenia now ran into various difficulties. In a still largely *laissez-faire* field of science with multiple taxonomic schemas (American Psychiatric Association, 1952), many authors would put forward their own interpretations. In doing so they could often gravitate towards the ideas of other thinkers, or indeed lay undue emphasis on certain tenets of Bleuler's theory.

For example, Bleuler clearly considered disease processes to underlie the group of schizophrenias: 'the group includes several diseases' (1911/1952, p. 8). However, he also speculated that an outside possibility existed that schizophrenia may have pure psychological origins. Travis (1924) then, who used the terms dementia praecox and schizophrenia interchangeably in the *Journal of Abnormal and Social Psychology*, would not have alarmed everyone with his formulation:

> Nearly all writers are agreed that dementia praecox is characterised by negativism, detachment of interest from the outer world, autoeroticism, expelling from the ego the impulses that have become unpleasant, shrinking of the ego, projection, and resistance to environmental factors (1924a, p. 292).

Yet this distinctly Freudian, and profoundly psychological reading, of the disorder once again overextends the particulars. Indeed, it carries the extension to its logical end: characterising the disorder solely in psychoanalytic terms. In some locations, schizophrenia was rapidly becoming much more psychological than Bleuler had ever intended.

Compounding this drift from Bleuler's core thinking, not everyone would accept Bleuler's nascent concept in sum or in part. Biologists, in particular, seem to have been slow to renounce the term dementia praecox (Kopeloff et al., 1935). So, for a time at least, the concepts of dementia praecox and schizophrenia ran in parallel and with crossover. Analogical bridging between the two concepts, in North America at least, was further made possible by a conception of dementia praecox, which emphasised—among other things—unconscious conflicts, psychogenetic forces, and splitting. In such a reading dementia praecox was also no longer quite Kraepelin's disease (Noll, 2011).

Both concepts remained subject to idiosyncratic interpretation. Psychologist James Winfred Bridges at Ohio State University, in his textbook *An Outline of Abnormal Psychology*, listed Kraepelin's clinical forms and declared them to share 14 symptoms in common. These included headaches and anorexia. Extreme emaciation from lack of food and headaches had been described earlier by Kraepelin (1913/1919). Bleuler agreed, calling anorexia an accessory symptom, but downplayed headaches. But symptom number 13 in the list supplied by Bridges was,

> Schizophrenia: or 'fragmentation of the psyche' (Bleuler) (1919, p. 77).

So for Bridges 'schizophrenia' was only a mere *symptom*. Not yet a disease or synonym to dementia praecox. A similar occurrence can be found in case files at Danvers State Hospital in 1918 (Noll, 2011).

Definition in flux

During the 1920s, in North America, schizophrenia achieved dominance over dementia praecox. A number of contributing factors have been identified. Bleuler's *Textbook of Psychiatry*, for example, was available in translation from 1924. Kraepelin's death in 1926 temporarily silenced an influence that had already long waned. And important organisations such as the Association for Research in Nervous and Mental Disease were now strongly signalling a preference for schizophrenia. Hence, schizophrenia was beginning to be acknowledged in classification manuals such as *The Statistical Manual for the Use of Institutions for the Insane* (Noll, 2011). By 1929, idiosyncratic interpretations of both terms or a confused blending of both Kraepelin and Bleuler ensured schizophrenia's definition remained in flux. As psychiatrist Charles Macfie Campbell of the Boston Psychopathic Hospital noted:

> The term schizophrenia as used in one clinic may bear with it assumptions and suggestions which it may not have in another clinic (1929/1934, p. 51).

[Privately, Campbell later expressed the belief in 1934 that there was no such thing as a disease schizophrenia. However, when 'angling for money' the term had a certain utility in acquiring funds (Noll, 2011, 2015).] Similarly, Jacob Kasanin and Moses Kaufman give four separate interpretations:

> Schizophrenia is a concept of many connotations. To a great group of psychiatrists the term still means the rigid definition of Kraepelin's dementia praecox. To others the fundamental reaction is the splitting and dissociation of the personality. To still others it is essentially the reaction of a maladapted individual type (A. Meyer). The psychoanalysts consider this syndrome a narcissistic regression psychosis (1929, p. 310).

Later researchers would further observe that not all workers even *within* the same clinic agreed on definitions of schizophrenia (Zubin, 1961).

Disentangling the Confusion

Given the ongoing flux, it is no surprise that in 1934 Howard C. Warren of Princeton University, in his *Dictionary of Psychology*, would attempt to disentangle the two ideas:

> Schizophenia = syn. For dementia praecox (Bleuler). (Literally 'splitting of the mind'. Schizophrenia has largely replaced dementia praecox in scientific usage. ... Somewhat broader than dementia praecox, since it includes, on the basis of dissociative symptoms, cases that would scarcely have received the earlier designation) (1934, p. 241).

Warren's exaggerated, but soon-to-be-true, declaration that dementia praecox had largely been replaced signals the fading of dementia praecox's influence. Nevertheless, schizophrenia did not stabilise. It continued to change. For instance, Isabella Wilson in *The Lancet* can be found informing readers that:

> schizophrenia, which included the form of illness known to some as dementia praecox, might present some of the following features ... a physique slender and 'asthenic' or dysplastic ... (1935, p. 327).

In doing so, Wilson, who would later cautiously assess insulin treatment, references Ernst Kretschmer's (1921) physiognomic analysis of schizophrenic body types. This idea linked body type (asthenic, athletic, pyknic) with disorder. And until critiqued in 1938 it was deemed to have some diagnostic utility and held sway over many psychiatrists (Farber, 1938). Wilson's account, as such, is a fossilised reminder against viewing a chronological series of psychiatric definitions as a process of adjustment and refinement. Indeed, in the DSM the asthenic personality would linger on until DSM-III.

Further Variations

Further variants and divergence in definitions continued to emerge. In 1938, for example, Despert observed that schizophrenia was a disease process but that it was accompanied by specific phenomena of dissociation and regression (1938, p. 366). Bleuler himself did *not* give any prominence to regression in his 1911 work. And, as such, this reading once again indicated the continued growth and competing influence of psychoanalytic thinking. This in itself was splitting into various strains by the mid-twentieth century. Of Despert's definition, Bradley's *Childhood Schizophrenia* observed that 'many workers would find no fault' (1941, p. 7). Yet by 1966 a critical Bührmann would observe that schizophrenia in childhood was an ill-defined syndrome, and that 'definitions have varied widely. Some have been so wide as to be meaningless' (1966, p. 921). Bührmann's own preferred criteria included rocking and failure to develop speech.

By now, attempts to differentiate dementia praecox and schizophrenia as distinct concepts had become increasingly futile. The confusion was seemingly ineluctable, and an attempt by Osborne (1940) to replace both terms with 'palaeophrenia' qua regression to a more primitive level fell on deaf ears. By 1940, psychiatrists Hinsie and Shatzky would note that schizophrenia was 'commonly synonymous with dementia praecox' (1940, p. 475). Note that these authors of the *Psychiatric Dictionary* were about to help reduce confusion by their elevation of the importance of 'cosmic identity' as a symptom. Cosmic identity—a fundamental reality to which all our reason must conform—can be traced to the philosophical writings of Constance Naden (1858–89). It was presented in a description of dementia praecox as something akin to a behavioural state. Hence, Hinsie and Shatzky wrote, 'The outstanding symptom of the simple form is withdrawal from reality; of the hebephrenic is cosmic identity; of the paranoid is persecution; of the catatonic is physical expression of negativism or positivism' (1940, p. 151). Notably, they also drew attention to 'morbid concepts' such as homosexuality. This formed 'the framework for the new and phantastic [sic] universe to which the patient adjusts himself' (1940, p. 150). They further isolated 'characteristics of primitive mentality', such as bisexuality. This definition of schizophrenia consequently functioned as both mirror and tool of social prejudice.

Definitions giving more emphasis to the various subtypes of the disorder now also emerged, but these served only to complicate matters. Richard Hutchings, Professor of Clinical Psychiatry at Syracuse University equated schizophrenia with dementia praecox and noted:

> dementia praecox (precox) (pree'koks) A psychosis usually appearing before middle life and characterized by introversion, repressed affect and interest. See hebephrenia; heboidophrenia; catatonia; paranoid dementia praecox (1945, p. 68).

Heboidophrenia [sic], a term derived from Karl Kahlbaum's heboides and rejected by Bleuler (Heuyer, 1974), was characterised as the simple type of dementia praecox. By this definition, Kraepelin's dementia praecox now served as rubric for a compromise mishmash of Bleulerian and Kraepelinian subtypes.

In some quarters, further facilitating both eclecticism and confusion was the continued belief that schizophrenia was not to be considered a unitary disease:

> The discussion on the definition and delimitation of the schizophrenias is not yet closed. Difficulties arose because some writers used the term schizophrenia as the name of a disease ... It is now more usually applied to a clinical syndrome (Curran and Guttman, 1946, p. 136).

In doing so Curran and Guttman (admitting that a dialogue of definitions was ongoing) sought to explain the disagreement over what exactly constitutes schizophrenia. Problems had supposedly arisen because some psychiatrists had neglected to consider the disorder a syndrome, which was now declared independent of cause, course, and pathology.

A Semantic Convention?

By 1946, R.G. Hoskins of Harvard Medical School, in *The Biology of Schizophrenia*, now posed the rhetorical question, 'what is "schizophrenia"? Is it an entity or, mayhap, merely a semantic convention?' (1946, p. 70). To this, Hoskins added:

> From the general biological point of view, however, it seems to me that the possibility must still be faced that 'schizophrenia' may be an entity by fiat only, as are disorders in general that are delimited merely on a basis of symptoms (1946, p. 72).

The possibility that the role played by semantics in the definition and concept of schizophrenia might deserve greater consideration complimented a number of similar critiques elsewhere. For example, in 1935 writing by Soviet G.E. Sukhareva notes the possibility that schizophrenia was simply a word that indicated lack of knowledge and not a word that corresponded to a real phenomenon (Zajicek, 2014). We will see more of such doubts later. Possibly underlying Hoskins' own particular doubts was the knowledge his research group had found it necessary, to utilise a fairly questionable subtype. This was 'the late indeterminate', a kind 'in which the characterising classificatory types have dropped out of the picture' (Hoskins, 1946, p. 91). Hoskins' observation appears to have met with silence. Nevertheless, it is important to remember that suggestions that schizophrenia was a semantic convention, or similar, predate the later doubts of post-1950s critics. Post-1950, similar comments would be treated as heresy. But they were not, in fact, new.

Dementia Praecox as Obsolete

In 1947, Philip Harriman's *Dictionary of Psychology* notes under dementia praecox that schizophrenia had by now largely, although still not completely, replaced Kraepelin's term. Nevertheless, the dictionary *rejected* any suggestion of synonymy; it gave different definitions for both schizophrenia and dementia praecox. Harriman includes a short history of Bleuler's victorious concept. In it, schizophrenia was, among other things, 'quickly adopted', over dementia praecox:

> Schizophrenia: Bleuler's term (1911) to describe a mental disorder characterised by autistic thinking. The term was quickly adopted in America as being more appropriate than dementia praecox (Kraepelin, 1883). ... Freud (1911) stated that it is the result of the unconscious homosexual trends; Boisen (1936) said that it is based upon conflicts pertaining to an intolerable loss of self-respect; many writers object to the psychogenic theories and uphold the view that it has an organic pathology (Harriman, 1947, p. 297).

Of course, elements of Kraepelin's thought could still be found in some corners of psychiatry. Bellak (1947/1952), for example, proposed dementia praecox qua biological could be distinguished from schizophrenia qua psychogenic (Noll, 2015). Nevertheless, by mid-century, in professional domains the *obsolescence* of dementia praecox was being definitively declared. And, indeed, irrespective of their detractors, psychogenic theories of mental illness—partially rooted in the work of Jung—had received considerable support from the surge of psychiatric casualties produced by war. R. MacDonald Ladell, in his *Dictionary of Psychological Terms* declared:

> Schizophrenia: A synonym for dementia praecox which has made the former term obsolete, and which has broader application ... It is a mental disorder often commencing in early youth characterised by an ever increasing fantasy life and a corresponding withdrawal of interest from the world of reality (1951, p. 42).

This declaration of obsolescence would have carried some weight as it was published by *The Psychologist Magazine*. Under the definition of dementia praecox, we find simply 'Obsolete see Schizophrenia' (Ladell, 1951, p. 11).

Post-1950s Definitions

By the 1950s schizophrenia continued to be defined with psychological parameters. For example, 1952 saw the release of the *Diagnostic and Statistical Manual for Mental Disorders,* otherwise known today as DSM-1. It characterised the schizophrenic reaction as falling under disorders of 'psychogenic origin, or without clearly defined tangible cause or structural change' (American Psychiatric Association, 1952, p. 5). As such, the conceptualisation of schizophrenia had moved a long way from the concepts of both Kraepelin and Bleuler, which emphasised and assumed a disease process. Nevertheless, it asserted the familiar claim to epistemological lineage—that schizophrenia was synonymous with the term dementia praecox. This shibboleth, or the contradictory alternative that dementia praecox was entirely obsolete, were now established readings of psychiatric history. And attempts to disentangle the two concepts from this point enter a period of stasis. [Scharfetter (1999) distinguished the two concepts, on constructs such as autism, loosening of association, the theory of primary and secondary symptoms, and the role of Freud. But one could go even further.]

Silvano Arieti in *Interpretation of Schizophrenia* now pressed the psychogenic theory further, and located the origin of the psychogenic reaction in childhood:

> At this point, we may again attempt a definition of schizophrenia, which takes into consideration the regression. Schizophrenia is a specific reaction to an extreme state of anxiety, originating in childhood, and reactivated later in life by psychological factors (1955, p. 384).

In the same text Arieti offered a remarkable re-reading of Bleuler's *Dementia Praecox or the Group of Schizophrenias* (finally translated into English in 1950). For Arieti, Bleuler 'delivered a blow to the Kraepelinian concept of Dementia Praecox as a disease entity' (1955, p. 14). Arieti was aware of the fact that Bleuler had spoken of a disease process. Yet he charged that Bleuler was unable to dismiss the possibility of an underlying organic process only because he could not explain everything with Freudian mechanisms. For Arieti, schizophrenia as a disease was now firmly rejected. Instead, he favoured a process whereby 'psychogenic factors may unchain a sequence of altered functionality of brain processes, involving even structures ...' (1955, p. 432). This was the high-water mark of psychoanalysis, and most department chairs of psychiatry in the 1960s were held by psychoanalysts (Decker, 2007). However, despite this hegemony, psychodynamic psychiatrists, preferring private office practices, were increasingly separated from asylum patients (Grob, 1994). And with this, there appeared to be little agreement over *which* psychological characteristics were the most important. The definition of the concept remained very psychological, very broad, and very open to interpretation. This also made biological research a formidable and extraordinarily difficult undertaking (ibid).

The Panchreston

As we shall see later, it was, in this climate of inexactness, interpretability, and loose definition that Thomas Szasz published his 1957 critique 'The Problem of Psychiatric Nosology' (Szasz, 1957). Szasz argued, for a number of reasons, chiefly concerning nosology, that schizophrenia had become an explain-all, a *panchreston*, filling a scientific void. Just like humours and protoplasm, schizophrenia explained little and, worse, obscured fundamental problems of psychiatry. As such, Szasz called for 'a conceptual clarification of the manifold meanings of the word ...' (1957, p. 409). Attempts at clarification were soon forthcoming.

Hempel's Contribution

In February 1959, the APA held a conference in New York on 'Problems in Field Studies in the Mental Disorders'. Problems of definition and taxonomy were moving to centre stage; Carl G. Hempel, a Princeton philosopher of science, had been asked to open the meeting (Zubin, 1961).

Hempel remarked that for a science to be objective it required that 'the terms used in formulating scientific statements have clearly specified meanings and be understood in the same sense by all those who use them' (Zubin, 1961, p. 5). He argued that to avoid deficiencies then prevalent in testing psychodynamic theories, *operational definitions* for scientific terms should be used. In championing operation definitions for schizophrenia, Hempel paralleled earlier comments by psychologist J.S. Beck: 'behaviours are our definitions. Schizophrenia is as schizophrenia does. It is out and out operational' (1954, p. 202). More specifically, Hempel was following P.W. Bridgman's (1927) operationism, which had earlier been heavily debated in the *Psychological Review* (Boring et al., 1945). Operational definitions allowed for objective criteria to be used in deciding whether a term should be used for a particular case. Elsewhere, operationalism had also passed from Bridgman's Harvard colleague Mandel Cohen to a trainee: DSM-III committee member Eli Robins (Decker, 2013).

Hempel chastised that the then used 'praecox feeling'—following H.C. Rümke in 1941 (Noll, 2015)—whereby many psychiatrists simply *felt* their subject was suffering from schizophrenia (or in the case of Binswanger, smelt). For Hempel, it did not constitute a worthy constituent of an operational definition. It lacked objectivity, although we can still find its use endorsed in French Psychiatry as late as 1975 (Ey et al., 1977).[2] Hempel may also have had in mind other subjective measures such as Southard's 1918 empathic index used for differential diagnosis in dementia praecox, along with 'past experience' and 'intuition' (Noll, 2015).

Searching for a Definition

Following Hempel's presentation, discussions about the meaning, use, and definition of schizophrenia abounded. In the process, the lack of agreement and clarity over the concept of schizophrenia was exposed in full. Professor of Psychiatry E. Stengel would laud Hempel, and remind psychiatrists that,

> A term should be employed in the same way by various observers. Here psychiatry leaves much to be desired. Take ... schizophrenia. This term is used far from uniformly (Zubin, 1961, p. 25).

Stengel had earlier complained in the *BMJ* of attending clinical conferences and having the uneasy feeling that *concepts* of schizophrenia differed considerably. 'It cannot even be said that although we are using different concepts of schizophrenia, we know a schizophrenic when we see one' (Stengel, 1957, p. 1176). (He further rejected 'process schizophrenia' and 'defect schizophrenia'.) Psychiatrist Leslie B. Hohman similarly complained: 'its meanings are so enormously different ... although we have the same word for it' (Zubin, 1961, p. 31). In an afternoon session, Danish psychiatrist Erik Strömgren declared, 'When, for example the word schizophrenia is used, everybody knows that the word is practically meaningless unless a detailed description is given of the sense in which the author wants to use the word' (Zubin, 1961, p. 173).

In yet another discussion, DSM committee member Moses M. Frohlich, with thoughts partially echoed by E. Eduardo Krapf (University of Buenos Aires), revealed an uncertainty very much hidden in the DSM classification and definition:

> The clarification of definitions ... is tremendously important. ... When schizophrenia is mentioned, it seems to evoke a different concept in everyone. ... This much difference in the concept of schizophrenia is incompatible with any kind of a common language, with

> any kind of common understanding, or any kind of camparable [sic] data (Zubin, 1961, p. 87).

Because of this difference, Frohlich later proposed to those in attendance that:

> we each define what we would put in the area of schizophrenia ... define the limits of these areas and circulate it among ourselves to see whether we agree at least on the definitions to start with ... (Zubin, 1961, p. 393).

Similar problems with differences in diagnosing schizophrenia, in teaching hospitals, and elsewhere were echoed by conference committee members Benjamin Pasamanick and Paul H. Hoch.

Harvard's Brian MacMahon presented an alternative way of understanding the role of definition in schizophrenia by noting, 'We are very intent on defining the case before we begin the investigation ...' (Zubin, 1961, p. 191). For MacMahon definition would be the outcome and endpoint of investigation. In contrast, Yale's Frederick, C. Redlich, argued that:

> we should try to start out with simple definitions and go to work, and in the case of schizophrenia it is very clear that we should stick to nuclear schizophrenias, the real schizophrenias, and leave out all the fringe enterprises of American psychiatry for the time being (Zubin, 1961, p. 203).

The idea of nuclear schizophrenia referenced by the Yale professor first appears in the work of Jacob Kasanin and Moses Kaufman in 1929—as a synonym for 'typical schizophrenia'.[3] However, it was quickly critiqued when E. von Domarus, in the presence of Kasanin, had asked:

> Dr. Kasanin speaks of nuclear cases; but here, so it seems to me, lies the central clinical difficulty. What are the central nuclear groups? Is there any symptom in schizophrenia that would be—all by itself—pathognostic? Or is there no such symptom, and does the diagnosis depend on a syndrome of symptoms? (Kasanin, 1933, p. 126)

By 1961, with such a critique unresolved, nuclear schizophrenia continued to be somewhat aspirational in nature. Oslo's Ørnulv Ødegaard argued that in psychiatry 'all definitions are more or less pragmatic.

This means they should be judged on their usefulness ... they should make communication and collaboration as easy as possible' (Zubin, 1961, p. 295).

The limitation of pragmatic definitions was, of course, their potential arbitrariness. In words that would appear to have later been seized upon critically by R.D. Laing (1967a), Brian MacMahon noted:

> Dr. Gruenberg, in private conversation I believe—not in the formal meeting—gave an operational definition of a schizophrenic as 'a person whom, after I have talked with him for 15 minutes I consider to be a schizophrenic' (Zubin, 1961, p. 334).

With uncertainty gaining the upper hand, Lothar B. Kalinowsky, staged a defence of the concept and argued:

> If there are differences as, for instance, in the incidence—supposed incidence—of schizophrenia. ... It is only that people unfortunately do not apply the definition properly, so that the differences in diagnosis seem enormous, and we confuse the statisticians completely (Zubin, 1961, p. 351).

Another defence was proposed by associate research scientist, E.I. Burdock who argued against overvaluing definition and reminded the conference that it was wrong to let lexicographers legislate meaning. In doing so, Burdock had touched on something important; perhaps definition *was* overvalued as a tool of psychiatric practice. This echoed the earlier thoughts of Hempel. He had stressed that while operational definitions may improve diagnosis they could not be 'generally considered as affording a definition of the concept in question' (Zubin, 1961, p. 42). DSM disorders, argued Hempel, *confused* definition and diagnosis. In other words, the assumption that definition was necessary for diagnosis was misplaced (irrespective of consensus). Although slow to spread, the repercussions of this international meeting were immense. Operational definitions, in the form of checklists with sometimes-optional criteria, would ultimately be introduced into or now be promoted within psychiatry. Definition would become unnecessary for the classification and diagnosis of schizophrenia.

While these ideas filtered slowly through psychiatry, debates over the definition of schizophrenia remained unexhausted and sometimes covered familiar ground. Moreover, not everyone would be happy with operational definitions. Epidemiologist and psychiatrist Ernest M.

Gruenberg later warned that 'some operational definitions are circular, imprison our thinking, and obscure issues' (cited in Gottesman and Shields, 1972, p. 12). Others could complain of 'measurement by fiat' (Ingleby, 1981, p. 31).

Antipsychiatry

While this process of reorientation and reconfiguration of definition continued, historically loose definitions of schizophrenia (frequently in conjunction with accusations concerning appalling patient treatment) remained under sustained attack. By 1961, Psychiatrist R.D. Laing would observe that many of the textbook signs of schizophrenia varied from hospital to hospital. And to him they seemed to be largely a function of nursing. Furthermore, some psychiatrists observed certain schizophrenic signs much less than others (Laing, 1960/1990). By 1967 Laing openly declared:

> Schizophrenia is the name for a condition that most psychiatrists ascribe to patients they call schizophrenic (1967a, p. 139).

Support for Laing's perspective on the social contexts involved in acquiring a diagnosis of schizophrenia came from self-proclaimed 'antipsychiatrist' David Cooper (although the term *antipsychiatrie* was earlier used by Bernhard Beyer in 1908). In *Psychiatry and Anti-psychiatry*, Cooper defined schizophrenia as:

> a micro-social crisis situation in which the acts and experience of a certain person are invalidated by others ... then confirmed (by a specifiable but highly arbitrary labelling process) in the identity 'schizophrenic patient' by medical or quasi-medical agents (1970, p. 16).

By microsocial Cooper meant a finite group of persons in face-to-face interaction. Laing and Cooper, each in their own way, and together, simultaneously proposed new ways of looking at individuals labelled with schizophrenia. Other contributions came from psychotherapists like Joseph Berke alongside the writing of patient Mary Barnes. Berke declared that he could never quite match the definition of the 'illness' [sic] with the reality of the people who were supposed to manifest it (Barnes and Berke, 1973). In doing so, all contributed to an atmosphere of renewed scrutiny over what exactly constituted schizophrenia. We will return to this later.

Improving Definition

One approach to growing self-doubt over definition, which remains as popular today as it is unsuccessful, was to search for biological markers to define schizophrenia. Often this was with the techniques of molecular biology:

> We have presented briefly some molecular biological concepts to help define schizophrenia because this would appear to be the pathway for the future (Teller and Denber, 1968, p. 110).

A second approach was simply to offer only a historical definition. Leland E. Hinsie and Robert Jean Campbell, in their *Psychiatric Dictionary*, now observed that, 'The concern about mental health issues that gripped the nation in the 1960s forced new perspectives in psychiatry' (1970, p. v). As such, they now provided eight pages of historical definitions concerning 'the schizophrenias'. These allowed the discerning psychiatrist to choose, in a schizophrenia à la carte fashion, just what was and was not to be defined as schizophrenia.

Thought Disorder

By 1968, the APA's DSM-II still studiously avoided the term disease. Instead, it characterised schizophrenia as 'a group of disorders manifested by characteristic thinking, mood and behaviour' (American Psychiatric Association, 1968, p. 33). Among other things, the DSM-II notably declared of the 'schizophrenias' that 'mental status is attributable primarily to a thought disorder ...' (ibid). This promotion of 'thought disorder' to the pantheon of core schizophrenic symptoms has a long history (Berrios, 1996). Thought disorder predates schizophrenia. It was, in a weak sense, reconceptualised by psychiatrists such as Bleuler into loosening of associations. After the decline in associationistic thinking, a 1927 paper by von Domarus (influenced by Vygotsky) marked its re-emergence in the schizophrenia literature. However, by 1975 Robert Spitzer would note tremendous disparity over what was considered thought disorder (ditto for fantastic or bizarre delusions) (Decker, 2013). Psychiatrist J.K. Wing would later argue, as would others, that it was not synonymous with schizophrenia. And that kinds of thought disorder occurred in many other disorders (Wing, 1978). In conjunction with a somewhat eclectic and elastic taxonomic schema, such ambiguity over thought disorder effectively meant that no consensus understanding of schizophrenia could be found through it. New definitions continued to be formulated.

The Need for a Heuristic Device

Throughout the 1970s the problem of the definition of schizophrenia remained extant. Mosher and Feinsilver in 1971 could speak of profound disagreement as to what constituted schizophrenia (Curran, 1974). For Reiss and Wyatt, the study of schizophrenia had been 'bedevilled by the problem of definition' (Reiss and Wyatt, 1975, p. 77). While Robert E. Kendell again complained, 'Unfortunately, although we all use the term schizophrenia there is no clear consensus about precisely what is implied' (1972/1975, p. 11). Indeed, the nature of the problem now seemed even wider. For Judith Greenberg there were no 'clearcut definitions of "non-schizophrenia" for in nosological language, there are definitions of pathology but none of normality' (1975, p. 11). For Beel, schizophrenia might even be partially described as a talent, 'in the sense that people who have it are able to notice things that others cannot' (Beel, 1975, p. 98). While, for Forrest, 'Schizophrenia in its widest definition may be more like a minority group than an illness' (1976, p. 291). Everyone, it seemed, was dissatisfied with the diagnostic systems of others. And many were unhappy with their own approaches (Strauss et al., 1974a).

Kendell, as such, seeking resolution, outlined a series of intranational and international discrepancies in diagnosis rates and argued in the spirit of Hempel:

> The only solution to these problems is to provide an operational definition of the term schizophrenia. ... It must provide precise rules of application enabling a firm decision to be made in every case on whether or not the criteria for the diagnosis are satisfied (1972/1975, p. 13).

Yet, even by now, not everyone agreed on the necessity for operational definitions. This could be seen even in seemingly 'harder' areas of core genetic research. Irving I. Gottesman and James Shields, in *Schizophrenia and Genetics*, clearly believing their own investigations were not hindered by a lack of definition, argued that,

> The syndrome of schizophrenia enjoys the status of an 'open concept' ... and need not be strictly defined operationally in order to retain its legitimacy as a concept (1972, p. 12).

But Gottesman and Shields were in the minority. The distinction between operational definitions and definition was not always clear to many authors. Yet the need to improve diagnosis was now a firm conviction. And for Salzinger it was clear that 'the general lack of precision in definition ... contributes substantially to the failings of diagnosis' (1973, p. 9). However, precision still depended on what you considered you ought to be measuring. Many, like Loren and Jean Chapman in *Disordered Thought in Schizophrenia*, reasserted this to be thought disorder, which was not without definition problems of its own:

> Most writers agree that the central defining symptom of schizophrenia is thought disorder. ... If one could describe the thought disorder and measure it, he might be better able to establish objective criteria for diagnosing schizophrenia (1973, pp. 11–12).

Nevertheless, the idea of thought disorder was gaining credibility. Theodore Lidz similarly noted:

> Schizophrenic patients suffer from serious disturbances of thought and communication. This is a matter of definition, for the presence of a thought disorder is the critical, though far from the sole, attribute of that category of psychiatric disorders we term 'Schizophrenic' (1978, p. 70).

Behind this thin veneer of consensus lay a grimmer truth. The debate over the semantics of schizophrenia had still not ended, and attacks by Szasz and others meant that the concept was still being vigorously reconceptualised. In 1973, an exasperated Kurt Salzinger, of the New York State Department of Mental Hygiene, could declare that

> schizophrenia is a unicorn. In neither case do we have definitive information about the cause of its appearance ... like the unicorn is described in various ways by various people (1973, pp. 1–2).

Moreover, he argued, whether or not these descriptions or the conditions for them existed, belief in their existence had significant consequences. Others, however, seemed less concerned: 'schizophrenia, like love' was 'a human condition that is recognizable but that defies unequivocal definition' (Jonas and Jonas, 1975, p. 35). Or perhaps it was a syndrome or a set of related psychiatric disorders (Reiss, 1975). All this could lead, for John Shershow at least, to a detached but revealing bemusement:

> I often think these days, as I used to in studying syphilis in medical school, that I could learn a lot from schizophrenia—if only I could find a case of it! The point, of course, is that the definition of 'schizophrenia' has varied tremendously throughout modern psychiatric history, not to mention the period prior to the modern era (1978, p. 4).

Refinement Through Consensus?

By 1980, critics such as Theodore R. Sarbin and James C. Mancuso were refusing to accept schizophrenia as a known or knowable entity. Instead, schizophrenia was a moral verdict masquerading as a medical diagnosis on norm violating behaviour (Sarbin and Mancuso, 1980). By contrast, for others some gloss had to be put on the instability of the definition. In the *Comprehensive Textbook of Psychiatry*, Herbert Weiner criticised the idea of a 'schizophrenia spectrum' as too broad. Yet he viewed the 1970s as a period of refinement through consensus:

> ... a time when many investigators are trying to refine its definition by developing a consensus along more restricted lines (Weiner, 1980, p. 1121).

Achieving social consensus appears to play an important role in ensuring the stability of concepts in science. Yet the idea of refinement through *consensus* appeared inherently fragile in this case. In the very *same* volume, Otto Allen Will Jr, would complain: 'The term "schizophrenia" is not easy to define; it has an elusive quality, and whatever is said about it is in some ways unsatisfactory ... schizophrenia as medically defined can properly be called the grave psychosis' (1980, p. 1217). Hogarty's earlier restatement of the problem of schizophrenia continued to reverberate, 'Whose definition of schizophrenia should we consider [?]' (1977, p. 587)

DSM-III

By 1980, DSM-III arrived with its long-gestating operational procedure for diagnosing schizophrenia (American Psychiatric Association, 1980).

For example, in diagnostic criteria A, one had to have displayed at least one of six possible signs/symptoms. In doing so, it introduced the 'schizophrenic disorders' and embraced a neo-Kraepelinian agenda (citing Schneider and influenced by growing genetic evidence—see later), which once more asserted the primacy of schizophrenia as a disease. Although DSM-III was institutionally accepted, dissenting conceptual viewpoints did not go away, as observable in a 1984 definition of schizophrenia given by Howells and Osborn. The definition largely corresponded to that of Bleuler, not DSM-III, and the authors noted that 'disagreements on aetiology, diagnosis, and classification are still widespread' (Howells and Osborn, 1984, p. 833). Numerous future revisions of schizophrenia in the manual would confirm such suspicions (we will see some examples later). One continuing source of dissent was that—even by now—not everyone favoured the idea of schizophrenia as a disease. For instance, Frank J. Bruno's *Dictionary of Key Words in Psychology* (1986) declared that schizophrenia was a functional disorder (i.e. no obvious pathology at the biological level). And yet other definitions continued the old tradition of attempting to integrate, standardise, or popularise new theoretical dispositions or ideas. For example, Mike Cardwell's definition in his *Dictionary of Psychology* introduced the increasingly fashionable idea of 'positive' and 'negative' symptoms (1986, pp. 204–5).

Pondering on how to choose between definitions, psychiatrist Ian Brockington would observe that there existed a 'Babel' of discordant definitions. And that, 'There is no use appealing to authority, since the leading authorities (Kraepelin, Bleuler, Schneider) disagree; that would make schizophrenia a social rather than scientific concept' (Brockington, 1985, p. 173). For Brockington, it was 'not easy to find suitable criteria for choosing between definitions' (ibid). For Brockington, improvement of methods and refinement of concepts were necessary. Possibly, a number of such concepts were on the table. For instance, the idea of relapse and preschizophrenia qua a prodromal period, or prodromal dementia praecox—traced to work by neurologist Charles Dana in 1904 (Noll, 2015)—were also described as slippery concepts and difficult to define around this time (Bower, 1985).

Different Meanings in Different Places

By 1987 Richard E. Kendell, Professor of Psychiatry at Edinburgh University, in the *Oxford Companion to the Mind* (Gregory, 1987), summed up, as he saw it, the problem of definition persisting among some progress in diagnosis:

> In the last decade, the adoption of unambiguous operational definitions, at least for research purposes, has reduced the confusion; though the coexistence of several alternative ways of defining the term still means that a diagnosis of schizophrenia may have a somewhat different meaning in different centres (1987, p. 698)

The observation by Kendell echoed that of the earlier mentioned Campbell as far back as 1929: 'The term schizophrenia as used in one clinic may bear with it assumptions and suggestions which it may not have in another clinic' (1929/1934, p. 51). Similarly, in 1987 Abou-Saleh would comment negatively on a neurodevelopmental hypothesis proposed by Murray and Lewis (1987). Abou-Saleh complained that it was uncertain as to what type of schizophrenia they referred to. This was in view of the fact that, for Abou-Saleh, schizophrenia was a heterogeneous disorder and that there were, '10 different available definitions' (1987, p. 1278). For Abou-Saleh, schizophrenia was a notional concept in the inner eye of the beholder, 'construed in a social context' (ibid).

Conclusion

This examination showed the historical eclipsing of the label dementia praecox. This was not so much by Bleuler's concept of schizophrenia, but by the rise of the label schizophrenia. This, in its own right, also created a superficial illusion of progress. However, other than serving as a vehicle for theoretical conjecture, progress as manifested through definition was slight. Instability in definition was not the exception. It was the norm. Often definitions were nothing more than a series of opinions about unknown facts, or generalisations on a descriptive level. In fact, only when consensus over the nature of schizophrenia eluded them did some investigators belatedly begin to look more closely at their terminology, assumptions, and modes of expression.

Perhaps unsurprisingly then, twentieth-century conceptualisation of schizophrenia in North America witnessed an epistemological shift from competing individual definitions to group-sanctioned operational definitions. In doing so, it abandoned an implicit reliance on definition—historically modelled on concept formation in geometry—as a way of representing or conceptualising schizophrenia. It moved towards a more explicit operationalism mixed with a convenient pragmatism. Yet for many, as the twentieth century drew to a close, the much-recognised problem of definition remained problematic.

Definition fluctuated. Nevertheless, it is important not to overstate the role of fluctuating definition in assessing the ontological status of schizophrenia. The schizophrenia of Bleuler may be attacked for failing to meet necessary and sufficient conditions (Boyle, 2001). Yet an open concept model of schizophrenia (Gottesman and Shields, 1972), or a DSM operational approach, often seems to sidestep such criticism. Each representation and interpretation of twentieth-century schizophrenia was historically constituted in a subtly different manner. And it is important that any assessment recognise the historical assumptions and limitations each embraced.

Clearly though, at least through *definition*, twentieth-century schizophrenia was not a stable transhistorical object. This supports and extends similar conclusions elsewhere (Berrios et al., 2003). Schizophrenia was frequently not, as Hacking (1999) might say, determined by the nature of things. It was not inevitable. And the twenty-first century could still be debating dementia praecox. Had there not been other problems with the conceptualisation of schizophrenia, we might brush all this aside. But, as we shall see, there were other problems. The variability of definition was in a sense symptomatic of a deeper malaise.

4
Catatonia: Faces in the Fire

It is impossible to examine all the historical subtypes of schizophrenia, with the kind of attention they might deserve. Nevertheless, we will attempt the study of one major subtype, whose problematic roots pre-date twentieth-century schizophrenia. Namely, we will examine a schizophrenia subtype, emerging from a concept earlier proposed by Karl Ludwig Kahlbaum in 1874: catatonia. And, as we shall see, as the twentieth century unfolded, descriptions of catatonia were increasingly accompanied by a narrative of disappearance. This will reveal much that is illustrative about the kind of problems incorporated into schizophrenia's conceptualisation. It will also reveal more of the contradictions, ambiguities, and inconsistencies upheld and tolerated by the concept's supporters.

For many, the quintessential person suffering from catatonia is one who remains frozen for long periods of time. There is some truth to this. Right from the beginning, Kahlbaum drew attention to immobility in the patients he would describe as possessing catatonia. And at the beginning of the twentieth century, asylums and their directors bore witness to it. Bleuler, for instance, noted that the severer cataleptic schizophrenics assumed 'definite attitudes for months and years' (1916/1924, p. 403). Another patient, Kraepelin noted, 'knelt for years on the same spot' (1913/1919, p. 145).

Yet when one reads the literature, it quickly becomes apparent that such postures were often seen in a different light. The symptom of *rigidity*, for example, was only one symptom of many in catatonia—and a minor one at that. Moreover, like cataleptic states, it was not always present. In fact, a separate symptom of catatonia described in the literature was *mania*, which implies something quite removed from rigidity. Yet another was *hyperkinesis*—whereby the patient appears in incessant motion. Where rigidity did exist, it was often short lived. The often transitory nature of the immobility was further clarified in a claim by Charles Féré. His patients remained frozen (*Figé*), for between 30 and 45 minutes, if not spoken to (Féré, 1890). This understanding of the often transitory nature of the immobility makes clear Stoddart's cryptic observation of a catatonic, 'He behaves as a statue, but he is not statuesque' (1909, p. 237).[1] Similarly, Bleuler could speak of a seemingly confused catatonic in an 'acute state' who surprised with his virtuosity as a chess player (1911/1952, p. 86). Hence, for the practising 1950s psychiatrist there was little ambiguity: 'even a beginner realizes a catatonic is not a paralyzed person' (Arieti, 1955, p. 238). Catatonia was always something more than just an absence of movement.

From the outset though, quite what catatonia was remained a matter of some debate. Kahlbaum did not restrict his concept of catatonia simply to issues of movement. In 1874, his first significant publication on the concept appeared: *Die Katatonie oder Das Spannungsirresen, eine Klinische Form Psychischer Krankheit* (katatonia or tension insanity, a clinical form of mental disease/illness). Through definition, Kahlbaum explicitly defined katatonie as follows:

> Katatonia is a brain disease of cyclical changing course, with sequential mental symptoms of melancholy, mania, stupor, confusion and finally idiocy (*Die Katatonie ist eine Gehirnkrankheit mit cyclisch wechselndem Verlauf, bei der die psychischen Symptome der Reihe nach das Bild der Melancholie, der Manie, der Stupescenz, der Verwirrtheit und schliesslich des Blödsinns* ...) (1874/1973, p. 83).

Symptoms included mobility problems, sensory disturbances, negativism (active and passive opposition), hallucination, and delusions. They also included food refusal, bizarre habits, verbigeration, flight of ideas, a fondness for diminutives, and arrested thoughts. In this variety of appearances, Kahlbaum saw unity, even if many of the symptoms were not always present. And in the neologism 'katatonie' he produced a new class of people, typically referred to as catatonics. The definition was often repeated in the literature. But, by contrast, and tellingly as it turns out, the case studies that accompanied Kahlbaum's definition were widely recognised as heterogeneous and much less likely to be cited (Berrios et al., 2003).

Rejecting Katatonia

Earlier, on 15 March 1872, Kahlbaum had presented his nascent katatonie at a meeting of psychiatrists in Berlin. There appears to be no record of what he said, but a brief and anonymous report in *Allgemeine Zeitschrift für Psychiatrie und Psychisch-Gerichtliche Medizin* concerning Kahlbaum's presentation is unfavourable (Anon, 1873). It suggests that his new concept—to be published in more detail later—was dismissed by Reimer of Sachsenberg as nothing other than *melancholia attonita*. Fraenkel of Dessau added further objections. He argued that pellagra could equally result in the vociferation that occurred in Kahlbaum's cases. In his defence, Kahlbaum emphasised the concept's prognostic features (most likely that of a sometimes favourable outcome). However, he probably weakened his case by calling attention to the most theatrical, but specious, features of his newly formed 'katatoniker'. These were the symptoms manifested by participants in mass outbreaks of perceived insanity by various 'convulsants' and those subject to the 'Swedish preaching illness'. Kahlbaum did not attend the next meeting in August (Anon, 1873), preferring instead to finish his monograph (c. September 1873). In doing so, he dealt comprehensively with the more serious of the two objections. Of katatonia's three declared variants, *mitis*, *gravis*, and *protracta*, Kahlbaum explicitly equated *melancholia attonita* with his most typical subtype *katatonia mitis*, which was accompanied by stupor and melancholia (*gravis* was characterised by melancholia and mania, while *protracta* designated cases often accompanied by late-developing nerve/muscle contraction and remission or intermissions).

The immediate reaction to the publication of Kahlbaum's 1874 work was nevertheless similarly unfavourable. Subsequent psychiatric narratives would forget nearly all dissenting voices. Orthodox thinking on catatonia, for example, suggests that following Kahlbaum's presentation most French, European, and American authors 'confirmed Kahlbaum's descriptions' (Fink and Taylor, 2003, p. 6). Such writing suggests that the concept itself was widely accepted. Yet rejection of the concept, either in sum or part, was both widespread and sustained. By 1877, Caspar Brosius observed that the concept had already been challenged several times (Brosius, 1877). And a decade later, Clemens Neisser reaffirmed that a number of 'outstanding' psychiatrists either had reservations or were hostile to Kahlbaum's disease (Neisser, 1887). In doing so, Neisser cites annual gatherings of German psychiatrists in Nürberg in 1877. He similarly cites meetings in Eisenach in 1880, as well as a gathering of psychiatric associations in Berlin in 1885. In the 1880 meeting, for example, Emanuel Ernst Mendel, W. Sander, and Franz von Rinecker (a teacher of Kraepelin) all denounced the concept (Kraam and Phillips, 2012). Mendel thought its existence groundless, while Sander and Rinecker felt it deleterious for the classification of mental illness. The concept of hebephrenia was having an equally torrid time of it. Most new editions of psychiatric textbooks mentioned hebephrenia, only to denounce it as nonexistent. Or they sought to incorporate it into existing concepts (ibid).

Rejection of katatonia took various forms. In 1878, Tigges simply declared that he could not agree with Kahlbaum's schema. In 1880, Dr von Reinecker and Dr Sander of Dalldorf also rejected the concept and expressed their opposition to placing katatonia in a list of separate diseases. While more notably in 1878, Westphal argued that he did not consider the manifestations, described by Kahlbaum, to be specific nor to be such that they were entitled to a place in a clinical group (Peterson, 1897).

Albert Behr later reported that Westphal gave a *famous* talk in Hamburg diagnosing Kahlbaum's cases as *Verrückte,* which translates well as 'deranged' (Behr, 1891). Although Kahlbaum had, describing a prisoner, used the term *partiell-wahnsinniger Verrücktheit* (partial delusional derangement) he had not systematically used the term in forming his concept. Indeed, he referenced *partielle Verrücktheit* as a different disease form. As such, we find the term *einen deutlich verrückten Charakter* (a clearly deranged character) applied to only one other patient, who, as it happened, was transferred to Görlitz by Westphal. The distinction is important because in the eventual acceptance of the concept, the term *katatonische Verrücktheit* would gain some currency. For example, A. Leppmann (1890) would prefer to 'subordinate' *katatonie* under *akuten Verrücktheit.*[2]

Others similarly preferred to place Kahlbaum's katatonie in their own favoured taxonomic boxes. In 1885, for example, Eugen Konrád in Vienna argued that cases similar to one that he had earlier presented could be found diagnosed in other asylums as katatonie (or melancholia attonita). However, for Konrád they more properly should have been considered as cases of *hallucinatorischen Verworrenheit* (hallucinatory turbidity or fogginess) (Konrád, 1885). In 1889, W.B. Lewis, in a *Textbook of Mental Diseases,* could argue that, having studied katatonia closely, he had become convinced that he was not dealing with any distinct pathological entity. Instead, he suspected multiple forms of hysteria (Lewis, 1889).

In 1892, Edwin Goodall wrote a critique of katatonia in the *Journal of Mental Science*. The critique claimed, with perhaps only a little exaggeration, to sum up the British reaction towards the concept. There had never been, Goodall wrote, any widespread enthusiasm about katatonia in Britain. For medical men in asylums it was either of 'doubtful significance' or 'without meaning'. He also pointed out that in Bethlem Hospital, cases claimed by Hammond to be katatonia had, in fact, received a different diagnosis. This held true even when there was some correspondence with the descriptions of katatonia in the psychiatric literature (Goodall, 1892).

At most, wrote Goodall, it seems to have appealed to individuals who have expressed merely *isolated views*. In doing so, he accused Julius Mickle of making statements about symptoms, which were 'singularly vague'. But even if katatonia could be found, he doubted it would be as frequent in England as in France, German or Austria–Hungary. This, Goodall claimed, was on account that hysteria (which, for Goodall, seemed a prominent feature of katatonia) was less common in England. Goodall further added that even in so-called cases of katatonia he had never observed the symptom of verbigeration (Goodall, 1892, p. 229). Perhaps it was a peculiarity of 'continental katatonia' (although French writers, he believed, had denied that verbigeration was characteristic of *any* disorder). Having left little of the concept of katatonia intact, Goodall concluded his paper by filleting Kahlbaum's pathological observations. Goodall, who cites German papers, appears to have a good understanding of the relevant literature of the period. But he simply did not accept the concept of katatonia. Nor, he believed, did his colleagues.

Such arguments undermined Kahlbaum's claim for an independent disease. In 1892, Dr R. Percy Smith, in the *Journal of Mental Science*, cited a case of what he dismissively described as 'so-called katatonia'. The fact that advocates of katatonia as a special form of mental disease did not hesitate to speak of katatonic symptoms in *other* varieties of insanity was just one reason against the use of the dubious term introduced by Kahlbaum (Smith, 1892). In the same year, at the *College of Physicians* in Dublin, Thomas Drapes also articulated the concept's rejection. Katatonia was too ill-defined to be considered as a distinct type (Anon, 1892). In 1895, L.W. Dodson, writing in the *Medical Record*, similarly stated that it was impossible to draw a sharp line between so-called katatonia and stuperous melancholia with cataleptoid symptoms (Freeman, 1895).[3]

Others found problems with the supposed course of the disease. In 1895, John Warnock of Peckham House Asylum (London) wrote that he had never met a case of insanity that followed the course described by Neisser (who accepted it) and other writers. As such, he was reluctant to diagnose his case with catalepsy as one of katatonia (Warnock, 1895). And perhaps for such a reason even Kahlbaum's former assistant Theodor Ziehen had rejected the postulated idea that katatonie had a cyclical course (Ziehen, 1892). Following a literature review in 1897, Peterson and Langdon (North America) declared that katatonia was not a distinct form of insanity. It was not a clinical entity: it was simply a form of melancholia. They noted, 'It is not desirable therefore to retain the name Katatonia' (Peterson and Langdon, 1898, p. 298).

Such reviews appear to be symbolic of a broad lack of acceptance of catatonia among many nineteenth-century alienists. If anything, these grew more entrenched as time passed. In 1902, for example, William Ireland, in the *Journal of Mental Science*, would declare:

> In my opinion, katatony is a formal distinction into which it is difficult to squeeze a sufficient number of cases of insanity. To find katatony one must hold Kahlbaum's description in mind, and step into the asylum to seek for examples. It is like looking for faces in the fire (1902, p. 582).

And as late as 1906, Thomas Drapes would argue:

> I ask anyone to read with an unbiased mind the descriptions given by different authorities of the so-called 'varieties' of insanity designated by the terms 'katatonia' and 'dementia praecox', and say in all honesty whether he has found any mental enlightenment therein, or whether he has not rather found himself reduced to a condition of intellectual bewilderment, more or less (1906, p. 79).

What can we say about this forgotten genealogy of dissent? First, such dissent reveals that the validity of Katatonie had been questioned from its inception, over a wide geographical area. Second, we must consider the length of time over which it occurred: some three decades. This is not particularly unusual in itself for debates concerning concept validity. But it does dismiss any idea that Kahlbaum's concept was quickly accepted. Finally, it indicates that even when accepted by some notable authorities, the concept was still not universally accepted. It remained contentious for quite some time.

As the century came to a close, the literature of dissent had grown so strong, that reviews of the dissent had begun to occupy numerous pages in communications dealing with the subject. This further suggests that a large number of alienists must surely have rejected the concept without ever formally publishing their objections. There clearly existed alternative and competing explanations for katatonie.

Acceptance

In certain quarters, however, as noted above, the concept did gain acceptance. Most importantly of all, acceptance occurred within pedagogical works that would influence a new generation of trainee psychiatrists. Notably, in 1880, Heinrich Schüle of Illenau lauded Kahlbaum in his *Handbuch der Geisteskrankheiten*. Schüle gave extensive treatment to the concept, and under *cerebropsychosen* classified *Katatonische Verrücktheit*. Schüle described this katatonic *Verrücktheit*—the term earlier emphasised by Westphal—as a hebephrenia with an associated motor tension neurosis (*Spannungsneurose*). Schüle was unable to confirm suspicions that such cases had a special disposition to tuberculosis. Katatonic *Verrücktheit* is indexed as *Katatonische Form der pubischen Verrücktheit*.

Schüle's thinking spread internationally but does not appear to have remained static. By 1886, the French translation of Schüle had placed 'catatonie' under a form of acute systematised delirium with hallucinations. The work also detailed depressive, expansive, and hysteric subtypes (Schüle, 1888). Hence, in this case *Katatonie* was accepted by Schüle only through modification. The distinction did not always pass without observation. In 1886, Charles Folsom had earlier equated katatonia with both the Katatonie of Kahlbaum and the Katatonische Verrücktheit of Schüle. He did so without drawing distinction (Folsom, 1886). But, by 1887, Clemens Neisser could point out that when one spoke of katatonie, one had to choose between the katatonie of Schüle *or* the katatonie of Kahlbaum (Neisser, 1887). In any case, Schüle's work was a textbook on clinical psychiatry and the first textbook to consider the concept. So the book's inclusion of the term (it has its own section in the contents) was significant. It represented a critical transition away from Katatonie being perceived as a highly debateable theoretical concept. Instead, Katatonie had now acquired the status of established object of pedagogy, perhaps in need of fine tuning. [Schüle himself remained wary of Kahlbaum's concept. In 1898, Schüle is noted as arguing, in his *Study of the Katatonia Question*, that there was no clinical entity that could, with justice, be called katatonia (Jelliffe, 1898).]

A subsequent, but equally important, event for the survival of katatonia was its further inclusion in Emil Kraepelin's *Textbook of Psychiatry*. This direct rival to Schüle's textbook ran through a number of successful editions. The concept of Katatonie first made the second edition in 1887. It was later associated with the Latin term dementia praecox (*dementia praecox* had already been used by Schüle, under the influence of Morel, as early as 1886). Kraepelin's influential textbook included most well-known psychiatric disorders. And on account of its perceived authoritative status, it was translated into English early in the twentieth century. By contrast, Kahlbaum's contentious monograph had to wait 100 years before it was translated. Without such pedagogic support, it is quite conceivable that Kahlbaum's Katatonie would not have survived in German psychiatric culture (never mind Anglophone). Its reception beforehand was tepid, its status precarious. After such inclusion, its status would be much more assured.

Kraepelin's support should be contextualised. In the broader scope of his thinking, Kraepelin encountered the logic and the methods of clinical psychopathology as proposed by Kahlbaum. Typically, this approach is abbreviated to a systematic use of the concept of *time* in facilitating diagnosis, prognosis, and so forth (Noll, 2007). Kahlbaum's method may not have been novel, Morel had something similar (Meyer, 1925/1928), but for many it was. Early on, J.J. Kerbert had credited Kahlbaum's katatonia with subsuming a number of independent diseases that many observers had supposedly *erroneously* identified as separate diseases. This occurred on account of the fact that *each individual stage* of the disease could last a long time (Brown, 1880). And others, such as Hecker had quickly recognised and boasted about the utility of the methodology in respect to katatonia.[4] So for Kraepelin, who like Kahlbaum shared an active desire to reform psychiatric nosology, this was obviously a methodology that could help reorganise psychiatric categories. And the early use of the method had been associated with the identification of katatonia.

Like Kahlbaum, Kraepelin further rejected the existence of a single brain disorder (*Einheitspsychose*). The latter had been supported by, among others, Heinrich Neumann (1814–88) with the aid of the idea of 'ametamorphose'. For Neumann, *Ametamorphose* derived from Esquirol's *lypémanie*, stupidity (*Stumpfsinn*), part of *mélancholie avec stupor*, 'Welt' *melancholie*, and, finally, the ecstasy of the writer (Neumann, 1859). By contrast, Kahlbaum's Katatonie was identified in the literature as a concept that not only rivalled, but was also superior to Neuman's ametamorphoses (Brown, 1880). For Kraepelin, accepting

Katatonie, over ametamorphose, would simultaneously attack the Einheitspsychose.

Yet Kraepelin (1904/2002) stated that he knew that Kahlbaum's ideas had long been contested. He also knew that Kahlbaum's cases were not homogeneous (Berrios et al., 2003). In fact, the heterogeneity of cases in Kahlbaum's Katatonie is so remarkable that it has been argued that Kraepelin's grouping of Katatonie with dementia simplex, paranoides, and hebephrenia was simply an act of faith (Berrios et al., 2003). The inclusion of a concept in a rival's textbook could not have been sufficient justification for accepting a contentious concept, for someone of Kraepelin's calibre. Nor does it seem likely that Kraepelin had simply been trying to attack an opposing viewpoint. Nor that he was trying to preserve the integrity of a methodology. So why then did Kraepelin accept Kahlbaum's concept in his 1887 *Psychiatrie*?

First, it is worth observing that the question carries an assumption. It assumes Kraepelin did accept Katatonie. This is not quite correct: like Schüle, he, too, modified it. Kraepelin also rejected Kahlbaum's tripartite division of the disorder. Instead, Kraepelin declared typical catatonia closer to mania states in catatonia, which Kahlbaum had also previously described. As such, Kraepelin grouped together 'those cases in which the *conjunction of peculiar excitement with catatonic* stupor dominates the clinical picture' (1913/1919, p. 133, original emphasis). Kraepelin (1887) could not accept Kahlbaum's assessment of the course and prognosis of the disease occurring across multiple conditions. And, indeed, he would only accept cases as, 'special, quickly passing forms of dementia praecox' (Kraepelin, 1904/2002, p. 32). His *Lectures on Clinical Psychiatry* also demonstrate that he was more fatalistic than Kahlbaum concerning recovery in patients diagnosed with the disorder. Kraepelin wrote, 'the patient is feeble minded and will always remain so' (1904/2002, p. 58). (Similarly, in 1896, he had argued that the condition was usually a permanent disorder (Kraepelin, 1896).) Kraepelin also thought Kahlbaum's description of catatonia 'in a certain direction too narrow but in another as too wide' (1913/1919, p. 132). But he wasn't more specific. Kraepelin (1913/1919) further explicitly included ideas of sin, whereas Kahlbaum explicitly excluded them. Kraepelin's acceptance as such was a qualified acceptance. An 1896 review of Kraepelin's work by Meyer described Kraepelin's domain of katatonia as being much broader than that of other alienists (Noll, 2011).

Kraepelin's continued acceptance of the concept—over various communications—appears to have found maintenance and reinforcement in the works of others. Kraepelin, for example, was further exposed to Kahlbaum's *Katatonie* through the supervision of Albert Behr's 1891 thesis *'Die Frage der "Katatonie" oder des Irreseins mit Spannung'*. Similarly, a small but growing number of supporting references appeared in the research literature (by authors such as Brosius, Arndt, Aschaffenburg, and Neisser). These may have alleviated Kraepelin's concerns over the concept. Certainly, Kraepelin increasingly cited them in later editions (whereas he initially only cited Kahlbaum). Such acceptance may not have lasted. On a visit to Java in 1904, he found only a modified form of katatonia. And it is thought that by 1920 he further called into question his formulation of dementia praecox (Berrios and Kraam, 2002). By this point, however, Kraepelin's thinking had already influenced Bleuler's conception of schizophrenia, which, like dementia praecox, had incorporated catatonia as a subtype.

Disappearance

In sketching his particular understanding of catatonia, Bleuler described in detail many curious symptoms. Examples included *catalepsy*, *stupor*, *hyperkinesis*, *stereotypy*, *mannerisms*, *negativism*, *command-automatism* and *echopraxia*, *automatism*, and *impulsiveness*. Bleuler diagnosed catatonia frequently and stated that:

> More than half of the institutionalized schizophrenics show catatonic symptoms, either transitorily or permanently (1911/1952, p. 180).

Unsurprisingly then, with more than half of his institutionalised patients showing catatonic symptoms, Bleuler considered catatonia one of the fundamental subtypes of schizophrenia.

In principle, all this could have led to a healthy comparative debate over the merits of catatonia as described by Kahlbaum, Schüle, Kraepelin, and Bleuler, and so on. However, it did not. Instead, later schizophrenia researchers soon encountered a significant problem that undermined the significance of any such potential debate. Numerous supposed cases of catatonia could still be found throughout the twentieth century. Yet for many it seemed catatonia was increasingly disappearing.

After acknowledging an initial rise in numbers (c.f. Jones in Bruce and Peebles, 1903, p. 625), the twentieth-century psychiatric literature announces the apparent disappearance of catatonia in a number of complementary ways. As early as 1919, for example, with the concept of schizophrenia only in its second decade, Bleuler's one-time assistant Carl Jung had stated:

> I must call special attention to the fact that the worst catatonic states and the most complete dementias are in many cases products of the lunatic asylum, brought on by the psychological influence of the milieu, and by no means always by a destructive process independent of external conditions (1919/1972, p. 215).

For Jung at this time then, catatonia could already be viewed as simply the product of the lunatic asylum and the psychological influence of the milieu. Subsequently, in 1939, with electroconvulsive therapy (ECT) only just introduced, and neuroleptics years away, Jung would speak of the,

> enormous change the average lunatic asylum has undergone in my lifetime: the whole desperate crowd of utterly degenerate catatonics has practically disappeared, on account of the mere fact that they were given something to do (1939/1972, p. 247).

When Jung speaks of crowding, he may have had in mind conditions at Burghölzli. In a 1907 letter to Freud he described it as being once again in a period of fearful overcrowding (Jung, 1974). It is also useful to mention that Jung also noted improvements in old catatonics, after transfer to new surroundings. And that he attributed this to psychological factors (Jung, 1907/1972). It is also important to note that Jung links the disappearance of catatonia to the phrase 'in my lifetime'. This is because mass deaths in German asylums during the First World War would have collapsed figures for catatonia in various locations (as later in French asylums in the Second World War and indeed the holocaust) (Aly, 1994; Von Bueltzingsloewen, 2007). Jung's observation also negates later claims, as made in a 1981 *Psychological Medicine* editorial 'Where Have all the Catatonics Gone?', that the disappearance of catatonia was due to the efficacy of antipsychotic drugs (the drugs were introduced after Jung's comments) (Fink and Taylor, 2003, p. 10). Jung also gave no credence to the efficacy of sporadically reported successes in treatment of catatonia via contemporary approaches such as barbiturate administration, whose effect appears to have been transitory (Bleckwenn, 1930). Consequently, we have a psychiatrist who had worked alongside Eugen Bleuler, during the formulation of the concept of schizophrenia, with its attendant emphasis on catatonia, who now firmly declared that such cases had practically disappeared.

Jung's use of 'practically disappeared' of course correctly implies that some cases could still be found. In 1936, for example, H. Dagand, writing in *Encéphale,* could declare that pure catatonia, as originally described by Kahlbaum in 1874 had little by little been forgotten. Why? Because of its relative *rarity* in the clinic (Dagand, 1936, p. 296). As such, Dagand could make much of his discovery of a case of 'intermittent catatonia' found in a Marseille asylum (St Pierre de Marseille).

Patient B, from Bastia, Corsica, had been detained almost continuously in his cell for over 20 years. Having initially presented with a history of suicide attempts, he had long since been considered beyond hope. He could now be found lying naked in his cell, in a fetal position. His intermittent catatonia (longest duration 4 months, shortest 15 days) was accompanied by a slight weakening of the intellect (*un affaiblissement intellectual léger*). Dagand admitted this was unsurprising for someone who has been continuously incarcerated for 20 years. Dagand's motives were neither therapeutically orientated nor aimed at explaining the rarity of catatonia. Rather, he had come to document and photograph his naked subject extensively. All this was with a view to capturing and affirming the purity of the concept of 'intermittent catatonia'.

Patient B was not so keen. Perhaps sensing this disappearance of his own subjectivity in this process of objectification, access was brusquely curtailed. B himself called for his nurses and ordered his own disappearance, that is, to be returned to his cell. Nevertheless, Dagand had by this time achieved his aims. As such, he managed to write up and publish his photographs of this exciting case of catatonia, which was both *pure* and *rare.*

Such photographs were not insignificant; the phenomenology of catatonia was intrinsically photogenic (in contrast to delusions or hallucinations). And, as reports of the rarity of cases of catatonia in schizophrenia increased, graphically frozen patients such as patient B would continue to be well, if not indeed disproportionately, represented in the iconography of twentieth-century psychiatric textbooks discussing schizophrenia or dementia praecox. In doing so, the all-fixating gaze of the camera could freeze a patient in a pose forever, as, for example, in Figure 30 of Bleuler's 1916 textbook—even if the pose was actually only assumed when the physician *appeared* (Bleuler, 1916/1924, p. 403).

However supposedly rare, plenty of patients could still be found elsewhere. And subtypes continued to be formulated. Hence, H.K. Stauder, for example, could describe *tödliche katatonie* or fatal catatonia in three catatonic cases who had hastened their deaths by repeatedly slamming themselves into the ground or walls (Fink and Taylor, 2003; Stauder, 1934).[5]

Yet regional differences could be striking. For example, an analysis by James May of 10,000 cases of schizophrenia in New York showed catatonia accounted for 12% of cases. However, variance in figures attributed to differences in diagnostic criteria showed rates of 23% in some institutions and just 7% in others. For some, such as Vivian Fisher, this failure in diagnostic uniformity could be rationalised. For Fisher, strictly speaking, such figures could only be ascertained by ignoring the mixed type. The mixed type, 'in a broad sense would include all cases' (Fisher, 1937, p. 342). Elsewhere, Gjessing argued that 'periodic catatonia' accounted for 'scarcely more that 2–3% of schizophrenics' (1938, p. 608).

People classified as catatonic were often thought to have been patients of long-duration illness (Quastel and Wales, 1938). Yet even they were not beyond therapeutic intervention. By 1938 in North America, even the neglected back wards were being reimagined as spaces of therapeutic possibility. John Romano, recalling conditions in chronic dementia praecox wards around this time, remembered patients had cyanotic feet. They also had 'blue piano legs' from 'sitting still all day'. Romano further recounted painstaking attempts at soliciting communication in patients who had seemingly not articulated sound in 20 years. These attempts were successful, although they were subject to relapse when treatment was withdrawn (Romano, 1977). That is not to say that conditions for schizophrenics had improved everywhere around this time. In 1938, a surprised Aubrey Lewis could report that in Waldu where 'Lutz ... is studying juvenile schizophrenia' mechanical restraint was still used 'not infrequently' (Angel et al., 2003, p. 92). But even cases of 'profound catatonic stupor' would respond if they could be convinced that the environment was not necessarily hostile (Biddle, 1949).

Whatever might be said about existing patients, however, it seemed that fewer and fewer new cases of schizophrenia were being diagnosed with catatonia. Around mid-century in the USA, the apparent disappearance of catatonia continued to be registered. In 1945, for example, David Rapaport could report—without apparent controversy—that cases of clear-cut catatonic psychoses were extremely rare. In his study of diagnostic psychological testing, only one case appeared over 'several years' (Rapaport, 1945).

Nevertheless, mid-century catatonic cases continued to be integral to the conceptualisation of schizophrenia. In 1956, for example, Dr Kinross-Wright would declare:

> I think there probably is a definite group of schizophrenic patients. Everyone would agree that if you go to any back ward of any state

> hospital you can find people who have been standing on one leg for many years. There aren't too many of these people, but I think that everyone would agree that they are schizophrenics (Bennet, 1956, p. 415).

As such, 20 years after Dagand's publication, new catatonic patients seemed to remain rare. And, in a sense, they were disappearing into an archetypal memory of schizophrenia: the frozen catatonic, who inhabited back wards. [Coincidentally, or not, this supposed practice of patient's standing on one foot, echoes a popular *fashion* dating at least as far back as 1865. This was a fashion where people aspired to ape the pose of classical statues (Root, 1865, p. 99).] In the already troubled conceptualisation of schizophrenia, such collective memories now constituted a definite group of schizophrenic patients.

Increasingly, although previously diagnosable as schizophrenia, many of the newer cases of so-called catatonia were considered to belong to other conditions. A 1937 report by Kleist and Dreist (1937), for example, found that 23.9% of patients diagnosed with catatonia were actually experiencing psychosis of 'feeblemindedness' or symptomatic psychoses. The latter had been precipitated by various infections, kidney disease, and thyrotoxicosis. Similarly, accounts of the misdiagnosis of catatonia in Wilson's disease and pseudosclerosis could be found in the literature (Bellak, 1947/1952). As such, by 1955, a defensive French schizophrenia authority Henri Ey would provide a cautionary summary of cases of 'pseudo catatonie' in various conditions. Ey lists malaria, syphilis, rheumatism, tuberculosis, cerebral sclerosis, cranial trauma, typhoid infections, and numerous other conditions such as *Escherichia coli* infections and alcohol poisoning. For all that, experimental catatonia, as induced in animals, would help shore up Ey's belief in a catatonia that was core to schizophrenia. Ey's belief held even if, by his own admission, the bad reputation of catatonic phenomena in the psychiatric clinic as being something *other* was largely *justified* (Ey, 1996). Elsewhere Redlich (1952) could similarly believe in the existence of an 'experimental schizophrenia' produced by drugs such as bulbocapnine, which induced stupor.

Yet, in fact, a continued, if not an accelerated, cultural shift in catatonia diagnosis seems to have been occurring in the 1950s. A quantitative 1969 study by Pauleikhoff discerned that changing diagnostic styles and procedures in clinical administration had led to a noticeable drop in the frequency of diagnosis post-1953 (Fink & Taylor, 2003). (Possibly, this then represents a second drop in frequency following earlier declines.)

By 1959, Marvin K. Opler can be found observing that certain forms of catatonia—the catalepsies that were once common in urban hospitals—were by then difficult to find. Opler put this down to historical changes in the *environment* occurring *without* changes in the organic substrate (although 'perhaps' catatonia remained higher in nonliterate societies). Opler (1959) confessed that to the orthodox organist such a declaration was worse than Galileo's heresy. But he did not renounce it.

One particular set of environmental variables was by now well understood. The 1960 edition of the influential textbook *Clinical Psychiatry* would more directly assert that in *well-run* hospitals catatonic stupor or excitement lasting for years was *rare*. In doing so, it drew emphasis to the need for patient's activities to be organised (Mayer-Gross et al., 1960, p. 264). And by 1968, Wing, drawing on Acheté and Ødegard, would speak of the almost total disappearance of catatonic stereotypies. These stereotypies had supposedly constituted a characteristic sign of chronic schizophrenia. Furthermore, negative symptoms were amenable to social therapy (Wing, 1968).

It should be noted that 'chronic schizophrenia' itself would later be criticised by the 1972 editors of the *Schizophrenia Bulletin*. For the editors it was a term with little descriptive or predictive value. They queried it as an iatrogenic condition resulting from treatment, lack of it, or treatment delivered late or not at all (Anon, 1972b). Similarly, the *BMJ* had noted 'compelling' evidence 'that many of the symptoms shown by the chronic schizophrenic patient are not an essential part of the disease process but are secondary and entirely preventable ... long-term stay in a hierarchically structured mental hospital exposes the patient to a high risk of damaging effects' (Anon, 1965, p. 141). Essentially, psychiatry had transformed the people it sought to explain. And patient apathy was no longer seen as a symptom of schizophrenia but instead as 'suitably adaptive' (Ortega, 1974, p. 5). It was now also recognised that staff, too, could be made chronic by poor conditions (Ortega, 1974). This does not necessarily make it simple to recast much of catatonia as an institutional issue. Certainly, one of Kahlbaum's case studies, a prisoner, had been so neglected that even his name had been forgotten. But institutional neglect varied historically and geographically. Jameson (1985) recommended the creation of a Michelin guide for institutions.

We are exploring a narrative of disappearance here, but the supposed disappearance of catatonia must not be overstated. In Florida, Monroe County records for 1960–66, for example, also showed the diagnosis was rare in some institutions. However, it was used consistently in others (typically between 5% and 10% of cases) (Guggenheim and Babigian, 1974). Similarly, in the summers of 1968–72 in Ponoka, Alberta, the number of vividly remembered cases of catatonia appears to have been very small but not absent (J. Martin, personal communication, 24 April 2015). In this sense, one finds a kind of Schrödinger's catatonia—the concept being simultaneously alive in one location and dead in another. That said, when a patient visited multiple facilities, he or she had less than a 10% chance of having an *agreed upon* diagnosis of catatonia (Guggenheim and Babigian, 1974). And for some the general overall trend was one of declining diagnosis of catatonia (Romano, 1977). Moreover, in 1977 Hogarty would further complain that patients who satisfied criteria for catatonia in one system of diagnosis were as likely to be diagnosed as nonschizophrenic in another (Hogarty, 1977). Interestingly, however, a statistical mean across institutions would populate some asylums with catatonia, even though none existed.

Post-1970, documented cases of successful treatment by pharmaceutical, ECT, and other methods, can readily be found. These continued a long, if sporadic, and variable treatment literature stretching back to Kahlbaum himself. It was now also recognised that drugs used to treat mental patients could actually produce catatonic states. In 1965, for example, we find an adverse reaction to an antipsychotic Mellaril (with an antidepressive called Elavil) characterised by catatonia, severe extrapyramidal involvement, acute renal changes, and suggestive blood changes. Thus, the very drugs being used to treat schizophrenia were capable of producing catatonia. And this was of sufficient concern for a clinical alert to be issued (Anon, 1965).

Despite the availability of various treatment options, speculation regarding the decline continued to occupy a notable place in the discussion of catatonia and hence conceptualisation of schizophrenia. For instance, in 1972, influential schizophrenia researcher Robert Kendell noted that catatonic *symptoms* had become steadily less common during the past 50 years. Kendell (1972/1975) thought this was as a result of broad social changes rather than because of any therapeutic advance. The allusion to broad *social* changes was doubtless true to some extent. Certainly, the pathologising of expressive religiosity and everyday sexual behaviour, as found in some early case accounts of catatonia (c.f. Kahlbaum himself), was increasingly out of fashion. The absence was seemingly not restricted to North America. In 1975, for example, catatonic forms of schizophrenia were also reported to have declined in Tunisia (Ey et al., 1977), although political events cloud any definite analysis here.

Speculation, however, remained uncertain and nebulous. In 1977, John Romano would feel that there definitely had been more patients diagnosed with catatonia in the early part of his career. His work had started around 1932. Romano (1977) felt that schizophrenic illnesses were now milder. Similarly in 1977, Murphy would argue that catatonia was much more common in the Western World in the nineteenth century. Murphy ventured that this may have been due to changing fashions in recording (i.e. diagnosis) or to unspecified changes that had occurred over time, regardless of culture (Murphy, 1977).

By 1977, the *Schizophrenia Bulletin* would also acknowledge that many clinicians had reported the virtual disappearance of the catatonic subtype (Durell and Katz, 1977). And when the twentieth century began its close this narrative of disappearance persisted in official psychiatry. In 1987, the American Psychiatric Association would similarly and uncontroversially declare that occurrences of catatonia were *rare*, while retaining the diagnosis within the DSM series (American Psychiatric Association, 1987). By 1994, DSM-IV's sourcebook explained DSM-IV's continued inclusion of the now rare catatonia. The apologetic authors complained that precious little literature existed to inform or guide their considerations about catatonic schizophrenia. To be precise, only one reference could seemingly be found, the patients from which were in India (McGlashan and Fenton, 1994, p. 436).[6] Nevertheless, because of its utility as a 'symptom complex', the authors argued that catatonia ought to be retained as part of the concept of schizophrenia (ibid).

Faces in the Fire

Late twentieth-century psychiatry continued to register catatonia as a component of schizophrenia. Its symptoms and startling iconography were retained in textbooks. Yet as the century ended, twentieth-century catatonia increasingly appeared to have become a ghost in the machine. Its chief characteristic was increasingly absence. It had become a once-prominent component of schizophrenia that refused to show itself.

There is no simple answer as to why this was the case, nor does there need to be. It is clear that there was nothing new in catatonia's problematic status. In an intellectual sense, the concept had been rejected from birth by many. The concept may have found a foster home in the crowded asylum of Bleuler's Burghölzli. It found a bed in the taxonomy of dementia praecox and schizophrenia, but it had always been heavily contested, modified, or denied. And this widespread critical doubt foreshadowed all subsequent rejection and disappearance of the concept.

For some, looking for catatonia had been 'like looking for faces in the fire' (Ireland, 1902, p. 582).

On another level, professionals were often reluctant or unwilling to diagnose the phenomenology of so-called catatonic cases as belonging to schizophrenia. The dismissal of 'pseudo catatonie' indicates that multiple other types of diagnosis, frequently of a medical nature, were probably favoured. On yet another level, many professionals also thought of catatonia as something passé. For many, it was the product of overcrowded and inhospitable asylums. As a result, being given something to do was suspected of resulting in positive outcomes. And perhaps under certain circumstances to diagnose catatonia was to indict one's own institution.

Variance in diagnostic criteria, itself indicative of the looseness in which both the concept of schizophrenia and the concept of catatonia were originally constructed, was also cited as problematic. Even when a unified constellation of pathological signs and symptoms were positively identified at catatonia qua schizophrenia, diagnosis in one institution did not guarantee diagnosis in another. Symptoms themselves were also reported as rarer and milder. Some now thought certain symptoms were no longer indicative of underlying disease. And, for some, unspecified social changes, not just therapeutic advances, were suspected as causal factors.

Catatonia's conceptual instability did not cause sleepless nights. For twentieth-century psychiatry, the increasingly problematic nature of the concept of catatonia was not out of place when housed within the concept of schizophrenia. Its disappearance was tolerated. Indeed, when read against a background of ever-changing definition it could not have been ranked as a particularly pressing problem. There were after all *some* catatonia cases. Yet to further understand how such instability could be tolerated and accepted within the concept of schizophrenia, it helps to look more closely and more generally at schizophrenia classification itself. For as it transpires, catatonia's instability was a subset of a larger problem faced by schizophrenia taxonomy. And as we shall see, there existed a host of more pressing contradictions, challenges and obstacles that faced those who sought to conceptualise schizophrenia. Only in the next century, via DSM-5, would the subtype of catatonia be removed from the concept of schizophrenia entirely.

5
Chasing the Phantom: Classification

Having examined catatonia, we now turn our attention to the broader classification of schizophrenia. This investigation, again spanning much of the twentieth century, will not just throw greater light on catatonia. It will complement, parallel, and at times further contextualise our earlier attempts to form an epistemological understanding of schizophrenia through our exploration of definition, the metaphor of splitting, and the schizophrenia lexicon. In doing so, this investigation will necessarily revisit the thinking of authors and spaces of theorisation we have already encountered, but without duplication. By building up such layers of understanding, we take another step towards a more comprehensive history of schizophrenia. We also resist premature simplification or synthesis of this complex concept.

It is natural for humans to describe, name, and classify. But the status of classification systems relating to mental disorder is an entirely different matter. In 1966, in *The Order of Things*, Michel Foucault asked what is the ground on which we are able to establish the validity of classification with complete certainty? And might not others classify similitude or difference in other ways? Most early twentieth-century psychiatrists did not tend to have such philosophical questions in mind when writing about schizophrenia or dementia praecox. And there appears to have been little explicit theorisation concerning either the appropriateness or the methodology of their endeavours.

Instead, at least until around 1954, as emphasised by Foucault in *Mental Illness and Psychopathology*, psychiatrists, like their counterparts in medicine, initially embraced prevailing derivations of earlier, Aristotelian-inspired, botanical taxonomy and classification, such as that found in the work of Thomas Sydenham (1624–89), Carl Linnaeus (1707–78), Boissier de Sauvages (1706–67), and William Cullen

(1710–90). For most, this was a way to order their findings, to establish their discipline as a science and to facilitate communication within it.

At times, biased by these existing traditions in psychiatry, the various twentieth-century subtypes in the dementia praecox and schizophrenia literature further came to be named as one might identify plants. And psychiatrists, when classifying patients, often legitimised their science using common botanical-like appendages. For example, *pseudo, mitis* (mild), and the suffix *oid* (meaning in the image of, or in the form of) are all used. In 1926, for example, Reiter—in deference to Kraepelin rather than Bleuler—reported on three patients with dementia praecox accompanied by motor symptoms. In doing so, he now appended 'parkinsonoides' to the term, which Nolan D.C. Lewis would, in turn, call a 'variety':

> As three of these patients had dementia precox with parkinsonian-like symptoms [sic] for years, Reiter thought they were a special variety of precox, i.e. 'dementia precox parkinsonoides' (1936, p. 114).

In contrast, and in deference to Bleuler, *Contemporary European Psychiatry*, edited by Leopold Bellak, gave the variety 'mild schizophrenia' or 'schizophrenia mitis' (Gylyarovsky, in Bellak, 1961, p. 286). The suffix *mitis* is readily found in plant taxonomy and indeed had also earlier been used elsewhere by Kraepelin and Kahlbaum (although the mild schizophrenia referenced here was a somewhat vague Soviet concept; for details see Zajicek, 2014). Similarly, we find talk later of pseudoschizophrenia (Langfeldt, 1937). As such, there was nothing particularly controversial when, in 1926, M.W. Boven from Lausanne could refer to dementia praecox in botanical terms as a 'species nova' (Bleuler and Claude, 1926/2001).

In parallel with such descriptive terminology, we naturally find the metaphor of the patient as a plant in the writing of researchers (although by no means all). Ernst Kretschmer, for example, found a clear biological affinity between body types and the psychic disposition of the 'schizophrene'. The latter included 'common or garden dementia praecox' (Kretschmer, 1925/1999, p. 15). Norman would state of dementia praecox, 'The field of dementia had remained a desert. ... I am confident that careful cultivation will show its fertility' (1904, p. 974). Similarly, Bleuler (1916/1924) could speak of a schizophrenic soil where manic and melancholic states develop. Louis J. Karnosh would write dedications to 'fellow workers engaged in the tedious tillage of the field of psychiatry' (1932, p. i). And Arieti, who would call Kraepelin the Linnaeus of psychiatry, would declare: 'The psychiatric hospital is a zoological garden, with many differentiated species' (1955, p. 12). Doubtless, such conceptualisation found its partial self-justification, inspiration and mirror in the large landscaped asylum lawns and vegetable gardens, peopled with patients and their problems.

There were, of course, alternative possibilities to classification. For example, for psychologist Lee Travis, the schizophrenoses and the psychoneuroses were two diametrically opposed groups. They existed on a continuum, the former characterised by negativism and restriction of reactions, and the latter by suggestibility and expansion of behaviour. The schizophrenoses (following Ernst Southard) included patients diagnosed as cases of dementia praecox, 'paranoiacs' [sic] and manic depressives. The psychoneuroses had a tendency towards hysteria (Travis, 1924, p. 297). But such exceptions breaking free from orthodox thinking were rare. Classification remained the measure of madness.

To facilitate our understanding of schizophrenia classification, this chapter begins with Kraepelin and Bleuler's formulations, which form the basis of subsequent attempts by everyone else to construct a classification of schizophrenia. The chapter then looks at the fluctuating states and uncertain boundaries of taxonomy in relation to their ideas. This extends our earlier understanding of problems concerning boundaries encountered in our exploration of definition. It also firmly helps us understand just how theoretical as opposed to clinically meaningful classification actually was. Next we look at some regional variations and briefly examine private classification. An understanding of variation in classification greatly magnifies and contextualises our earlier findings regarding varying definition, and further internationalises the concept's historical variance. Finally, it looks at other diagnosable categories in relation to schizophrenia. This is an important aspect of the history of schizophrenia that avoids examination if we simply look at definition, splitting, or the lexicon. All this contextualises the rise of the conceptualisation of schizophrenia through official classification. And, in particular, it will inform our later analysis of the DSM series that came to dominate twentieth-century psychiatry.

Kraepelin

In the fourth edition of Kraepelin's 1893 textbook, *Psychiatry a Textbook for Students and Doctors* (*Psychiatrie: Ein Lehrbuch für Studirende und Aerzte*), under 'Psychic Degeneration Processes' (*Die psychischen Entartungsprocesse*), Kraepelin separately indexes three conditions:

A. Dementia praecox: light and severe forms (Hebephrenie)
B. Katatonie
C. Dementia paranoides, depressive and expansive.

This was not yet what came to be seen as the classical formulation of dementia praecox. However, Kraepelin noted—rather ominously, as we shall see—that 'between these forms there are numerous transitions (*Zwischen diesen Formen giebt es zahlreiche Uebergänge*)' (1893, p. 435). Of dementia praecox, Kraepelin notes:

> with dementia praecox we describe the sub acute development, of a peculiar, simple, mental weakness in the young (*Als Dementia praecox bezeichnen wir die subacute Entwickelung eines eigenartigen, einfachen geistigen Schwächenzustandes im jugendlichen Alter*) (1893, p. 435, my translation).

The term 'démence précoce' had earlier been used as a descriptive term by Morel and other French authors (see also Ginguene (1794) for early idiosyncratic usage). Notably, Morel is cited by Heinrich Schüle when he uses the term dementia praecox in *Handbuch der Geisteskrankheiten* (1880). It now seems clear that Schüle heavily influenced Kraepelin's work (although the term dementia praecox can also be found in an 1882 piece by Van Deventer). So there exists a clear intellectual bridge from Kraepelin to Morel. Hence, in this spirit, Henri Claude (1926) would retrospectively diagnose one of Morel's patients, T. Séraphine, as possessing 'schizophrenic dementia' (*démence schizophénique*).[1] Kraepelin (1920/1922) also separated out manic depressive conditions but by 1920 would admit that he could not satisfactorily distinguish between the two diseases. Famously, Kraepelin used a system of cards or *Zählkarten* to order his patients scientifically, but recent studies have shown that they often contained scant information. Moreover, large numbers, maybe as much as half, lacked a diagnosis and information on the course of the illness (Decker, 2013)

Kraepelin's dementia praecox competed with the ideas of other prominent researchers, such as Erwin Stransky and Karl Wernicke (to name but two). However, the fifth edition of Kraepelin's textbook (1896) was broadly welcomed in North America. In a review not without criticism, Adolf Meyer, whose 'psychobiological' research has been described as more bureaucratic than biological (Noll, 2011), called attention to dementia praecox and spoke of 'new and revolutionary conceptions' (Meyer, 1896, p. 302). By 1899, in the sixth edition of *Psychiatrie*, Kraepelin classified catatonia, hebephrenia, and paranoid conditions under the rubric dementia praecox. And it would be the seventh edition (1903) from which Eugen Bleuler would cite and claim theoretical continuity.

Kraepelin was initially pessimistic about the chances of recovery in dementia praecox, but by the time of the seventh edition of his textbook his views had become more optimistic. Nevertheless, most psychiatrists continued to read him as a pessimist. For example, Dublin alienist John Conolly Norman, addressing the British Medical Association, protested that although Kraepelin's ideas had value, dementia praecox was not incurable (Norman, 1904). In the same manner, later psychiatrists and psychologists would cite the importance of Kraepelin. But they would continue to lambaste Kraepelin's choice of name—although often for their own rhetorical purposes. For others, Kraepelin's concept was already more problematic. For E.C. Runge in 1900, then superintendent of St. Louis Asylum, Kraepelin's concept merely functioned in a negative way. It acted as a default category for patients not easily classified. Hence, 'there are many cases which puzzle us, and we do not know how to classify them. I have found dementia praecox a very comfortable, I may say vulgarly, dumping ground' (Runge, cited in Noll, 2011, p. 103; Hill, 1900). By 1904, fears were being expressed by Bernard Sachs and C. Farrarr that dementia praecox was creating a pseudoepidemic (Noll, 2015). While, in 1916, William Alanson White would declare that 'Praecox is nothing more than a waste-basket into which we throw all the cases we know nothing about' (cited in Noll, 2015, p. 1)

Following Kraepelin's various editions, other more accepting researchers quickly extended Kraepelin's ideas to other areas. In doing so, Bleuler's assistant Otto Diem (Bleuler, 1911/1952), possibly following Sérieux (Heuyer, 1974), would describe in 1903, for example, the 'simple form' of dementia praecox. This would soon be adopted by both Bleuler and Kraepelin. Lloyd Andriezen would speak of 'dementia praecox criminalis' (Norman, 1904). And in 1905, Sante De Sanctis linked dementia praecox with childhood via a condition similar to catatonia. This appeared in very young children as 'demencia precocísima' or 'dementia praecox prepuberalis' (Lombardo and Foschi, 2008; Rojas et al., 1996). Although, as it happens, the latter was later dismissed by Sullivan: 'We encountered no case of recognized schizophrenic psychosis below the early or homosexual phase of adolescence: the so called *dementia praecoxisma* seems rather scarce' (1962, p. 165).[2]

Bleuler's Grouping

In 1911, Bleuler alternatively declared that 'for the time being dementia praecox must be regarded not as a species of disease but as a genus ...' (1911/1952, p. 279). In doing so, Bleuler's *Dementia Praecox or the Group of Schizophrenias* introduces the subgroups as:

1. The paranoid group
2. The hebephrenia group
3. The catatonic group
4. Schizophrenia simplex.

Schizophrenia simplex was hard to detect. It was usually found outside the hospital and accessory symptoms such as hallucinations and delusions were deemed absent. It included people 'who make our world uncertain' and could be recognised in labourers, pedlars, and servants at the lower levels of society. At the higher level of society, the most common type was the nagging wife and the eccentric. Bleuler also found it passing unrecognised in many alcoholics. Elsewhere, as with the earlier 1908 communication, Bleuler (1911/1952) mentioned a fifth subtype. This was the *latent schizophrenic*. Among other things, the latent could be definitively identified by 'social uselessness' (Bleuler, 1916/1924, p. 437). And according to Bleuler's son, Manfred, the use of 'schizoid' was also introduced into clinical discussions around 1910. This seemingly represented a mild variant of schizophrenia (Bleuler, 1972/1978).

Eugen Bleuler also noted that schizophrenia might be combined with other psychoses such as the oligophrenias (congenital mental defect, retardation, feeblemindedness). This could form cases of *Pfropfhebephrenia* (Bleuler, 1911/1952). This was also more generally known as engrafted schizophrenia or *Pfropfschizophrenia* (Bellak, 1947/1952, p. 426; Bleuler, 1916/1924, p. 437). Later, Bleuler would supplement and support his nomenclature with a more detailed exposition of the schizoid personality. And others, in turn, would supplement these terms in their own way. Ludwig Binswanger would, in *Schizophrenie*, for example, later extend schizophrenia simplex further. Hence, he could report with emphasis 'the polymorphic forms of schizophrenia simplex' (*der polymorphen Form Der Schizophrenia simplex*) (Binswanger, 1957, p. 179).

It should be noted that in adopting such a strategy Bleuler appears to have transgressed some basic taxonomy principles concerning the unique designation of a member of a taxonomic category. These principles demand that (1) there should be no synonyms; (2) no polysemy (more than one class of member per name); and (3) there should be no unnamed classes or members (Slaughter, 1982, p. 66). Bleuler violated (1) by declaring schizophrenia synonymous with dementia praecox. In fairness, Bleuler was not the only one to do this. Thomas Clouston for most of his career believed his concept of adolescent insanity to be synonymous with dementia praecox. This was despite significant differences between the two concepts (Ion and Beer, 2002). Nevertheless, Bleuler's blending of schizophrenia and dementia praecox would cause much confusion. Bleuler violated condition (2) by declaring schizophrenia to represent a group of schizophrenias. And, finally, to further muddy the waters, Bleuler declared the existence of unnamed members.

A second important observation is also worthy of note. In the 1911 text, Bleuler's subtypes—catatonia, paranoia, hebephrenia, simple schizophrenia, and latent schizophrenia—give the impression of being placed statically under a genus. But, in fact, they were *not* supposed to be invariant disease types. Bleuler makes clear that his subtypes were clinical *states* not separate diseases: 'All distinctions appear vague ... a closer study of our cases showed ... a complete absence of any distinguishable boundaries ... we know of no natural lines of demarcation within this group; what, up to now, has been considered as such boundaries are boundaries of clinical states, not of diseases' (1911/1950, p. 280). Bleuler's muddled distinction was recognised in 1912 when a favourable review by Van Teslaar, in the *Journal of Abnormal Psychology*, endorsed Bleuler's thinking:

> The condition ... partakes of the nature of a genus having varied clinical manifestations which frequently masquerade as different nosological entities (1912, p. 374).

For Van Teslaar, then, clinical manifestations were only masquerading as nosological entities. Later theorists would not be, for the most part, so astute in recognising this distinction. This is not surprising as it is difficult to spot without a very close reading of Bleuler's text—a reading that perhaps only a diligent reviewer like Van Teslaar would give. As such, where not ignored, the fact that Bleuler's subtypes were not actually disease entities went largely unnoticed or misread.

Recognisable tensions were lurking then within the conceptualisation of schizophrenia right from its inception. Yet for all that, Bleuler seemingly managed to square to his own satisfaction the boundary problems. He also squared the unsuccessful mapping of his subtypes with the reality of changing content and fluctuating form. How was such a feat possible? Bleuler did so by simply acknowledging that his analysis was provisional and wanting. He declared that 'the subdivision of the group of schizophrenias is a task for the future' (Bleuler, 1911/1952, p. 280). This may have seemed like an act of intellectual honesty. But it may be argued that in deferring itself to the future, Bleuler's schizophrenia was effectively announced through prophecy.

In 1913 Emil Kraepelin, in a move that negates all claims to a linear progression from dementia praecox to schizophrenia, responded to Bleuler's rival system. (This must be contextualised: the two were on fairly friendly terms. Kraepelin, on occasion, paid visits to Bleuler's home). Kraepelin's updated thinking (1913/1919) acknowledged that schizophrenia might come to challenge his label dementia praecox. However, he saw it as only one of a number of possible alternatives. Consequently, he did not faithfully embrace schizophrenia. Instead, Kraepelin indexed 11 subtypes, which fell under a lesser number of variant headings in the text itself. The heading 'paranoid dementias', for example, detailed the paranoid forms further divided into paranoid dementia gravis and paranoid dementia mitis. Such further division mimicked the exactness of the botanic taxonomy psychiatry idealised, but did not necessarily provide much in the way of additional diagnostic or conceptual clarity for those in clinical practice.[3] Kraepelin, as such, more or less lists:

A. Dementia praecox simplex
B. Silly dementia praecox; hebephrenia
C. Simple (depressive) dementia praecox stupor
D. Delusional depressive dementia praecox
E. Circular, periodic, and agitated dementia praecox (all belonging to 'the agitated dementias')
F. Catatonia; excitement, stupor (melancholia attonita)
G. Paranoid dementia praecox gravis
H. Paranoid dementia praecox mitis
I. Confusional speech dementia praecox, schizophasia.

Like Bleuler, Kraepelin would now also adopt Diem's dementia simplex. And with respect to the final form, 'confusional dementia praecox',

Kraepelin declared it to be a disorder of expression of speech. It was a disorder with little impairment of the remaining psychic activities. Consequently, he stated that 'If one will, one may therefore, relying on Bleuler's nomenclature, speak of a schizophasia' (Kraepelin, 1913/1919, p. 178).

In the same text, largely ignored in North America (Noll, 2011), Kraepelin would also outline the paraphrenias. This was a condition similar but contrasting to dementia praecox, where a loss of inner unity was limited to intellectual faculties. And of which Kraepelin could detail 'paraphrenia systematica', 'paraphrenia expansiva', 'paraphrenia confabulans' and 'paraphrenia phantastica'. Kraepelin was seemingly not an enthusiast for mixed forms. He had reportedly declared that an assumption of transition forms was merely an 'admission of nosological cowardice' (Meyer, 1910, p. 275).

Fluctuating States

In spite of his aversion to transition forms, Kraepelin would note, as before, that that there were 'numerous transitions' (1913/1919, p. 89). Kraepelin also noted 'that in spite of all efforts it appears impossible at present to delimit them sharply' (ibid). And he further noted, that 'the delimination of the different clinical pictures can only be accomplished artificially' (ibid). As such, he did not attribute special clinical value to his grouping (Kraepelin, 1913/1919).

All this meant that neither psychiatrist's classification of schizophrenia or dementia praecox was actually entirely isomorphic with taxonomic principles of botany. Despite the use of impressive scientific terminology, like genus and species, on close inspection, the classifications of Bleuler and Kraepelin were only loosely similar to these principles, although they may have believed otherwise. Instead, what Bleuler and Kraepelin offered as subtypes were fluctuating clinical states, or forms of schizophrenia or dementia praecox. Unlike varieties of plants, these could morph *into and out of* each other in short periods of time. Hence, in 1936, Nolan D.C. Lewis, speaking of catatonia and hebephrenic features, would complain that the two reactions were so frequently combined that he preferred to call this class of disorders the 'catatonic-hebephrenic group' (1936, p. 36). Bleuler's son, Manfred, gives a fair account of the still-extant problem over half a century later:

> During the course of years, one and the same patient often exhibited the most varied schizophrenic states. Catatonic, hebephrenic,

> and paranoid manifestations would follow one another in irregular succession and then interchange with states that correspond to the syndrome of schizophrenia simplex. ...To be sure, there are many patients who reveal no marked changes in schizophrenic symptoms over many years. By and large, however, one is impressed with the variability of the manifestations, more than with their consistency (1972/1978, p. 439).

The lingering presence of this problem over half a century is somewhat explainable. First, at times Bleuler and Kraepelin, in their enthusiasm for theoretical and taxonomic progressions, also appear in their writing to forget or mask this key observation of fluctuation. And as such, they perpetuated its concealment. This continued to occur even when, for example, Bleuler could admit that hebephrenia 'now constitutes the big trough into which are thrown the forms that cannot be classed with the other forms' (1916/1924, p. 426). Second, where fluctuation was explicitly noticed the tendency was to ignore it. Subtypes, wrote Vivian Ezra Fisher, could only be ascertained by 'Ignoring the mixed type, which in a broad sense would include all cases' (1937, p. 342). Dorcus and Shaffer, in their *Textbook of Abnormal Psychology*, explained—while blending the two concepts—why and how subtypes persisted as follows:

> While it is practically impossible to fit any of the cases definitely within a rigid classification, it is quite helpful for obvious reasons to have some scheme that will allow us to see the similarities of some cases. Following Kraepelin, then, we are able to distinguish four types of schizophrenia; simple, hebephrenic, paranoiac and catatonic, fully recognising the fact that most cases will be mixed in type (1934, p. 223).

Taxonomy, consequently, made visible to science, in a ceremonial space, categories of people who were not in fact there. And, indeed, sometimes the subtypes themselves were not even there. In 1911, for example, Bleuler had declared, that despite its apparent disruptive qualities, 'The simple type is hardly ever seen in hospitals ...' (1911/1950, p. 227). In the same spirit, he also asserted the existence of unnamed types without drawing negative attention. Likewise, in 1938, R. Gjessing, Director of Oslo Municipal Hospital, would leave an unjustified letter B for a possible type *not yet discovered*. This occurred in his 'synchronous–syntonic group' (the companion of an asynchronous–asyntonic group) and for Gjessing and the editors of *The Journal of Mental Science* was not, it seems, anything to be disturbed by (Gjessing, 1938, p. 613). In many ways, such a problematic nature was nevertheless hidden by various emerging and misleading reports in the literature of therapeutic efficacy regarding subtypes, via, for example, treatments such as insulin shock (James et al., 1937).

The Boundaries of Schizophrenia

In the century that followed Bleuler's 1908 paper, schizophrenia rehearsed and anticipated itself exhaustively. Each new classification that emerged simultaneously represented a critique and rejection of earlier forms. Yet simultaneously each classification established continuity with earlier traditions. That schizophrenia was broader than dementia praecox—as we saw when discussing definition—is readily agreed upon in the early historical literature. Hence the 1928 observation by a Soviet textbook noting Bleuler's significant expansion of the boundaries of dementia praecox is essentially correct (Zajicek, 2014). Even Bleuler himself had spoken enigmatically of schizophrenia as dementia praecox plus lighter forms of the illness (Bleuler and Claude, 1926/2001). Only later, did the two concepts come to be incorrectly seen as synonymous.

Owing to the indeterminate nature of schizophrenia's boundaries, taxonomy naturally suffered. Already by the 1930s Russian émigré Zilboorg would contemplate the problem of accounting for the so-called borderline cases. These were the 'incipient schizophrenias', 'schizoid maniacs', 'schizophrenics in manic phases' and 'psychoneuroses with affective episodes', and so forth (Zilboorg, 1931) Later, classification would extend schizophrenia's boundaries even further into such realms as the 'schizogenic family', as reported by B.B. Wolman in 1973. Wolman himself (1937) would also outline forms of aretic, autistic, symbiotic, and pseudoamentive childhood schizophrenia.

Historical categories such as 'other', 'unspecified type, 'dementia praecox not otherwise specified', and 'atypical schizophreniform psychosis' bear witness to a broader uncertainty. It should be stressed, however, that problems of diagnosis were not simply confined to fringe cases of schizophrenia—later officially referenced as borderline schizophrenia (Arieti, 1974a), and which Roy R. Grinker would speak of as a 'confusing waste-land of schizophrenia'—before subtyping it further (Grinker, 1971, p. 65).

Nor did anyone ever successfully identify a core to twentieth-century schizophrenia. Indeed, despite early attempts by figures such as Jacob Kasanin and Moses Kaufmann (Kasanin and Kaufman, 1929), no nuclear schizophrenia subtype ever found global acceptance in the next half century. No pathognomic symptoms were ever identified (we will see more of this later). That said, the idea of a nuclear schizophrenia continued to occupy a position in the theoretical literature as an idealised concept. Hence, as late as 1976, Martin Roth, working in conjunction with H. McClelland, argued that there was a continuing need for a strict *concept* of a 'nuclear schizophrenic syndrome'.

Roth and McClellan graded a series of disorders commencing with nuclear schizophrenia and extending through a hypothesised *spectrum* of psychoses: (1) nuclear schizophrenia; (2) paraphrenic and paranoiac psychosis, including 'late paraphrenia'; (3) 'cycloid' or 'episodic' schizophrenic (or schizoaffective) psychosis (antecedent psychological stresses absent or unimportant); (4) schizoaffective psychosis (without evidence of cyclical course or clear psychological stress); (5) 'psychogenic' schizoaffective or paranoid psychosis (onset impressively related to adverse life events); (6) toxic schizophreniform psychosis; (7) schizophreniform illness in association with cerebral disease (Roth, 1978, p. 72). Yet in recorded criticism, Psychiatrist Timothy Crow objected. The categories would not be mutually exclusive. Any one case could fit into a 'whole lot of your borderline categories' (Roth, 1978, p. 77).[4] Yes, Roth admitted. No sharp lines between nuclear schizophrenia and schizophreniform disorders, in the borderlands of the disorder, could yet be drawn. The system was premature (Roth, 1978).

Regional Concepts

Contributing to the problem of discerning core and boundaries was the fact that schizophrenia classification was also characterised by what we can call regionalism. In the case of North America, for example, a growing cultural confidence meant that it began to prefer its own formulations to Germanic variants. A similar parochial confidence can be seen within French and Soviet classification. This occurred even although twentieth-century debate on classification was often internationally informed.

In the Soviet Union, Lev Rozenshtein's idiosyncratic 'mild schizophrenia' briefly gained local currency around 1932. Mild schizophrenia could display various vague 'microsymptoms' or even symptoms that were mutually exclusive of schizophrenia. Although it was milder than schizophrenia on account of Soviet socialism, and although it was partially created in response to official pressures, it was later denounced by Soviet psychiatry. For this reason, in 1935, Victor Osipov could be found dismissing mild forms of 'schizophrenia without symptoms', complaining that surely one could not take this turn of phrase literally' (Osipov, cited in Zajicek, 2014. p. 189). Decades later, the Moscow Institute of Psychiatry would use a classification of 'chronic progressive schizophrenia' (comprising nuclear, paranoid, and slow progressive), 'mixed schizophrenia', and 'periodic schizophrenia' (circular, oneiroid, depressive, and acute paraphrenia). Although, like elsewhere, the Soviet Union itself also remained subject to internal regional variations (Beckett, 1971).

On account of such variation, one North American review would note that 'terminological and theoretical differences have prevented access of non-Soviet psychiatrists to the classification' (Holland and Shakhmatova-Pavlova, 1977, p. 277). Citing among its examples, the review included the fact that Soviet psychiatry had less emphasis on using early psychosexual history in diagnosis (Holland and Shakhmatova-Pavlova, 1977). However, aside from such periodic inspections, each locus appears to have privileged its own system of classification and modes of production and ontogenesis. Such systems were never stable or truly internationally isolated (although the Soviets came close). Yet enthusiasm for *importing* yet more contentious and competing classifications was low.

All of this, perhaps inevitably, led to huge differences in opinion as to what could be legitimately classified as schizophrenia. In 1938, for example, Aubrey Lewis made the following observation on schizophrenia diagnosis with respect to Moscow:

> Schizophrenia is a diagnosis more generally applied in Moscow than in Western Countries. Many cases we would call depressive, manic, or psych-neurotic they label as schizophrenia—especially schizophrenia mitis (Angela et al., 2003, p. 120).

Lewis appears to have had some justification for his numbers. Certainly, around this time, in at least one Moscow institution some 81% of patients were diagnosed with schizophrenia, with further claims being made that some 60% of schizophrenic cases remained undetected in the community (Zajicek, 2014). For this reason, even within Russia, complaints could be heard that psychiatry as a science had begun to turn into 'schizophren-ology'. [The situation was ultimately brought to an end by the Soviet government (Zajicek, 2014, p. 168).] Lewis would also state that one of the many difficulties in accepting Norwegian Rolv Gjessing's work was 'the uncertainty as to whether the cases he has studied are mainly schizophrenic as he suggests' (Angel et al.,

2003, p. 135). In particular, 'a number of manic-depressive, or what would be diagnosed as such in other countries, go into his statistics as schizophrenics' (ibid).

In North America, cases of 'pseudoneurotic schizophrenia' were first described by psychiatrists P.H. Hoch and P. Polatin (1949). They described a syndrome whose symptoms are not usually considered characteristic of schizophrenia. Patients presented 'pan-anxiety' and 'pan-neurosis'. This was an all-pervading anxiety structure in which gross hysterical mechanisms and vegetative dysfunctions such as poor sleep, anorexia, vomiting, palpitations, phobias, obsessions, and compulsions could all be present (Arieti, 1974b). Many cases of pseudoneurotic schizophrenia would have 'short psychotic episodes or later become frankly schizophrenic' (Hoch and Polatin, 1949). Hence, Stone and Dellis would speak of a test conducted on 20 hospitalised patients alternatively diagnosed with pseudoneurotic or 'pseudocharacterological schizophrenia' (1960, p. 333). Yet Kendell (1972/1975) later observed that the pseudoneurotic group was composed almost entirely of patients who would not be regarded as schizophrenic at all by most British psychiatrists. However, it should be re-stressed that such problems can be characterised much as one of regional discord as opposed to simply international discord. Kendell (1972/1975), for example, could also point out differences between New York and California (the latter using a narrower concept). Meanwhile, the concepts in St Louis, Missouri, and Britain were roughly similar. For Maurice Porot, Alain Couadau, and Bernard Aubin (1968) nosological criteria further varied according to school, although I cannot as yet confirm this generalisation.

In any case, international discord was not trivial. In 1955, a graphic illustration produced by Henri Ey clearly showed that, irrespective of its psychoanalytic component, the French concept of schizophrenia was much *narrower* than what he then perceived to be the global conception of schizophrenia. At the same time, American thought was described as an artificial ecological mythology based on infinite maladjustments, where the symptoms were banalised and made superficial (Ey, 1996). Hence, in 1976, Larousse's *Grande Encyclopédie* would continue to emphasise in its description of schizophrenia (drawn from Morel, Kraepelin, Chaslin, and Bleuler) that French psychiatry had always considered schizophrenia as something strictly defined. In contrast, Anglo–Saxons were claimed to have a large and excessively vague concept (Larousse, 1976).

French Historian Elisabeth Roudinesco would describe DSM-III as an attempt to 'liquidate the Freudo-Bleulerian nosology' (1986, p.486), and cites Marc Lander's declaration that DSM-III was an undeclared war against Europe. Yet when confronted with American hegemony (as made manifest in DSM-III), French researchers were quick to realise that publication in American journals would be heavily dependent on the use of the new nomenclature—irrespective of their acceptance of it. And European pharmaceutical companies recognised that in order to enter the American market, they must gain Food and Drug Administration approval—something that could only be done by using DSM criteria for patients in clinical drug trials (Pichot et al., 1983).

Quite how the French adapted to American hegemony remains to be historically analysed. By 1989, Kellam detected a growing convergence of opinion, but nevertheless he charted numerous differences. Hallucinations, for example, were not stressed in the French definition, 'presumably because if prominent they lead to a diagnosis of a chronic hallucinatory psychosis' (Kellam, 1989, p. 155). By contrast, 'derealisation and depersonalisation are not mentioned in the American definition, and they are not usually regarded as definitely psychotic phenomena in the Anglophone countries' (ibid). Bizarre delusions, and hallucinations commenting on the patient, were not given the same importance by the French, 'who do not seem to consider Scheiderian "First Rank" symptoms as important diagnostic characteristics of schizophrenia' (ibid). Moreover, French criteria demanded age of onset before 40 years, in contrast to DSM-III-R. And when compared with DSM-III-R, the duration of critical psychotic symptoms in DSM-III-R needed to be only one week, whereas in French criteria it needed two months. In his conclusions, Kellam deemed the potential acceptance of chronic hallucinatory states as separate from schizophrenia as unlikely in Anglophone psychiatry. As the century closed resolution had not been found. Commenting on French nosology in 1994, Pull and Chaillet would declare that French classification was essentially compatible with other traditional systems, particularly the International Classification of Diseases (ICD)-9, but that 'major peculiarities are to be found almost exclusively in the definition of schizophrenia' (1994, p. 24).

Understandably, then, in the individual researcher there often lurked a hesitating unease in the face of supposed scientific fact. This was hardly helped by peripheral texts that might describe something like simple schizophrenia as being 'characterised by a let-down in ambition, a lack of interest in occupation and an emotional indifference' (Wick, 1940, p. 1072). Hence, Romano could observe, 'I never was quite sure of how reliable or valid was the diagnosis of simple schizophrenia, and later on was even more dubious of the concept of latent schizophrenia and its sister subtypes, pseudoneurotic and pseudopsychopathic' (1977, p. 540). All this could sometimes lead to hapless attempts at resolution. By 1973, for example, Kohn would declare that he would use schizophrenia in the broad sense that it was employed in the USA, rather than a limited European sense. He did so not because it was superior. In fact, he thought it cruder. Instead, he did so because it had been used that way in much of the research that he would cite (Kohn, 1973).

Individual Classification

Individual researchers continued to produce idiosyncratic formulations alongside official manuals. Hans Maier (1922), for example, would identify and describe insurance hebephrenia (*versicherungshebephrenien*). In such a case, a patient's claim for insurance compensation would become more illogical and incoherent over time (Mayer-Gross et al., 1960, p. 272). [Bleuler himself would class 'litigious schizophrenics' (1916/1924, p. 414) under paranoid types.] Elsewhere, Kirston Weinberg, in the *American Sociological Review*, gave us the 'transient schizophrenic'. This was a person 'characterised by a relatively normal childhood and adolescent breakdown and a favourable chance for improvement or recovery' (Weinberg, 1950, p. 600). The 'transient schizophrenic' would stand in opposition to the 'chronic schizophrenic', and Weinberg claimed that the adoption of a dichotomy between chronic and transitory had other precedents, such as,

> the distinction between the chronic schizophrenic and the acute schizophreniform, as: endogenous vs. exogenous, constitutional or somatogenic vs. psychogenic, true vs. pseudo, predisposed vs. situational, classical vs. atypical, malignant vs. benign, process vs. episodic (ibid).

Unsurprisingly then, individual researchers could sometimes find themselves possessing both new ideas and the facilities to circumvent official taxonomy. Around 1935, for example, C.M. Campbell, during his later years as Director of the Boston Psychopathic Hospital, constructed his own private formulations. He had these typed out and distributed to the staff at the hospital—although he apparently never published a formal grouping in any journal. Campbell's formulation gave six unnamed groups, A–F. Details of this brief classification only emerged when provided by Samuel J. Beck in 1954.[5] Such private approaches still embraced classification itself. Nevertheless, their emergence constituted a rejection of official classification (as well as prior taxonomies and the so-called subtypes of others). Indeed, by 1975 Duchet would write that one could say that *everyone* had his or her own concept of the disease, and that the concept varied as a function of culture and era (Duchet, 1975).

Official manuals then often led a parallel existence sometimes at odds with theorisation by schizophrenia researchers. It should be remembered, that in day-to-day asylum practice things were also different. Early tabulation for diagnosis was limited. As late as 1913, for example, over 200 US public institutions for the insane did not include tables of patient diagnosis in their annual reports. This held even although they tabulated lots of other things (Noll, 2015). Early case books and dusty record books can be equally unrevealing when it comes to use of classification in diagnosis. And even when data were collected, their value remained subject to doubt. In 1917, Meyer wrote privately to Samuel Orton forcefully asserting that the statistics published annually were 'a dead loss to the States that pay for them, and an annual ceremony misdirecting the interests of the staff' (Meyer, cited in Grob, 1991, p. 426)

Reversible Schizophrenia

Complicating taxonomy further as the twentieth century unfolded was yet another problem. Schizophrenia researchers began to realise that a portion of their cases—the exact percentage is unclear—were not what they first seemed to be. In 1922, psychiatrist Karl Menninger introduced a number of new terms into the schizophrenia literature. Contemplating *postinfluenzal schizophrenia*, Menninger had deduced that dementia praecox was apparently sometimes a chronic delirium, or *schizophrenia deliriosa*. Conversely, delirium could sometimes be an acute dementia praecox or *delirium schizophrenoides*. This led Menninger to postulate a new form of schizophrenia that would subsume both these ideas. Menninger called it *reversible schizophrenia*. In his abstract, Menninger would emphasise,

> The concept 'Reversible Schizophrenia' (c.f. the term 'delirium schizophrenoids') is not a mere nosological quibble, since it may affect our conception of the nature of dementia praecox. It implies conditions of reversibility which we may discover to be under our control! (1922, p. 573).

By 27 March 1925, Menninger wrote a private letter to William Alanson White, the influential psychiatrist and translator of Bleuler's theory of *Schizophrenic Negativism*. In the letter, Menninger wrote:

> At the American Psychiatric association I am to present some more influenza studies this year in regard to schizophrenia. The upshot of the cases seems to be that most of what we thought dementia precox [sic], according to the old terminology and the old conception, failed to materialize as such, or at least after more or less schizophrenic fireworks lasting all the way from a month to a couple of years, the majority of them cleared up and the patients are as well as ever (Faulkner and Pruitt, 1988, p. 61).

Menninger contemplated: 'Such findings indicate either that our diagnoses were wrong or that our conceptions of dementia praecox are wrong, or that influenza produces a curious and atypical type of dementia precox [sic] which tends to recover' (ibid). Furthermore, he added, 'It has been my contention for several years, as you may or may not recall, that there is no essential difference from a psychopathological standpoint between simple delirium and dementia precox [sic] except chronicity' (ibid).

Bleuler did, of course, provide a differential diagnosis between schizophrenia and encephalitis in his *Textbook of Psychiatry*. Yet Menninger's experience was by no means unique. In 1953, A.S. Chistovich expressed a similar opinion regarding infection. Hence, the psychosis that occurred in connection with 'streptococcal, staphylococcal, colon bacillus infections, that is, with the ordinary "pyogenic" infections, correspond to the concept of "acute schizophrenia"' (Simson, 1960, pp. 452–3).[6] Christovich further believed that 'fatal' or 'very active' schizophrenia was an acute toxic infectious septic psychosis (Simson, 1960). [See also 'febrile schizophrenia' (Jaspers, 1962) and 'hypertoxic schizophrenia' (Fink and Taylor, 2003; Romasenko, 1953 cited in Hofmann, 1963).] In this reading, an infectious psychosis could be schizophrenia. And schizophrenia could be an infectious psychosis. Similarly, Alexandru Obrégia had introduced the novel 'syndrome schizoïdie de Régis' for cases following toxic infections after lactation or very big shocks (Bleuler and Claude, 1926/2001). While as late as 1971 we can still find reports of encephalitis being misdiagnosed or 'presenting as' as acute schizophrenia (Misra and Hay, 1971)

As it transpired, over the twentieth century a large number of natural kinds were variously diagnosed as schizophrenia. These would later come to be seen as supposedly misdiagnosed as schizophrenia. Yet although 'misdiagnosis' could certainly happen when a differential diagnosis was available, confusion also happened in many cases *before* a differential diagnosis was ever possible. That is to say, the differential diagnosis of certain conditions was identified and understood only long *after* the concept of schizophrenia was formulated.

The presence of epilepsy, for instance, is also noted within Kraepelin's early conceptualisation of dementia praecox. And indeed Bleuler had declared that 'many of our patients were first sent to us with the diagnosis of epilepsy' (1911/1952, p. 175). Bleuler had tried to separate it out, but the results were unconvincing. The necessary technological and conceptual advances to demarcate epilepsy clearly from schizophrenia, or other conditions, were not available in 1911. Moreover, even with such advances confusion continued. Hence, as late as 1959, epilepsy specialist H. Strauss would object to epileptic patients being put 'in the wastebasket diagnosis of schizophrenia' when there were positive electroencephalographic findings. And in 1974, N.R. Zec would similarly speak of a pseudoschizophrenic syndrome (Arieti, 1974b, p. 472). Most notably, a celebrated 1963 analysis of schizophrenia in four quadruplets, 'the Genain quadruplets' acknowledged but set aside the observation that epilepsy was present in at least two of the girls (and a diffuse toxic organic brain disorder in a third) (Rosenthal, 1963).[7]

Problems also emerged in relation to the concept of autism. As early as 1912, Bleuler had abandoned the word autism in favour of 'dereism' (Bleuler, 1912/1951). He felt dereism had fewer Freudian connotations. Yet few followed his lead. Most continued to use the term autism, and widely so. Indeed, the fundamental importance of autism, in must be noted, was often highlighted in midcentury schizophrenia research. Robert Volmat (1958) had even declared that autism constituted the pathognomic superstructure of schizophrenia. Yet, confusingly, autism would also become a powerful descriptive term for children (and later adults).

Notably, in 1943 it was used by Leo Kanner for 'a unique "syndrome" in children not hithertofore reported' (1943, p. 242). Yet Kanner's research on autism was derived from a small study of 11 children. Some of the children had previously been diagnosed as schizophrenic and possessed 'remarkable similarities' to schizophrenia (Kanner, 1943). How forcefully Kanner communicated the supposed distinction between these two conditions with 'remarkable similarities' is unclear. Later, in 1948, after interviewing Kanner on early infantile autism, *Time* magazine wrote of 'diaper-aged schizoids', or 'frosted children'.

These were children who 'had what Dr. Kanner calls "early infantile autism"; it is, he thinks, a diaper-age form of the mental disease called schizophrenia (split personality)' (*Time*, 1948, p. 2). Either way, confusion between 'infantile autism' and 'childhood schizophrenia' remained in the literature decades later (Gair, 1972). Psychiatrist J.K. Wing reported professional confusion between schizophrenia and autism as late as 1978.[8] And this, in a subfield where the absence of professional consensus in diagnostic terminology was later described as 'endemic to the field' (Gunderson et al., 1974, p. 33).

Frequently, the concepts of 'schizophrenia' and 'manic depressive' illnesses also overlapped. As early as 1920, Kraepelin (1920/1922) had expressed lack of confidence in an ability to distinguish between the two diagnoses. Eliot Slater similarly complained in 1936 that confusion between manic depressive illness and schizophrenia hampered his investigations (Decker, 2013). While, Bellak argued it was 'frequently all but a toss-up whether a case should be diagnosed schizophrenia with affective features or manic-depressive psychosis with schizophrenic features' (1948, p. 80). By 1978, Goplerud was still able to cite growing evidence that bipolar depressive illness was frequently misdiagnosed as acute schizophrenia (Goplerud, 1978). Some improvement appears to have occurred post-DSM-III. And, for Holden (1986), the recognition that some schizophrenia-like behaviours were actually attributable to mania (or other disorders) was not without consequence. It had actually contributed to a decline in the diagnosis of schizophrenia. But for most of the century, confusion between the two concepts was not considered surprising.

On rarer occasions, there also existed confusion surrounding some 'iatrogenic' conditions. In September 1946, for example, Dr Max Levin, writing in the *American Journal of Psychiatry*, would introduce the 'transitory schizophrenias' or 'bromide schizophrenia'. These were psychotic states produced by bromide intoxication, which had been a recognised side effect of medical use of bromide for decades. Bromide schizophrenia was declared to be one of four possible varieties of bromide psychosis. It was considered similar to 'ordinary' paranoid schizophrenia. But it was nonetheless 'accompanied by disorientation, a symptom which does not belong to the schizophrenic picture' (Levin, 1946, p. 229). It was thought to occur in people with strong schizoid leanings, and bring to the surface a latent schizophrenia. As such, this was a novel schizophrenia conceived as a brain *state* but actually acquired from the toxicity of medical *treatment*. [A similar argument may be applied to the occurrence of so-called "malignant catatonia" associated with the administration of antipsychotic drugs (Fink and Taylor, 2003).] But many cases were simply confusing or difficult. And they were written up to alert others. For example, Estella M. Hughes could report in 1925 that certain cases of Huntington's chorea had been observed to show a schizophrenic picture (Hughes, 1925). While for Lewis and Minski, 'the differential diagnosis between a chorea psychosis and a schizophrenia' was 'at times difficult' (1935, p. 538).

By 1978, there remained a wide acceptance that schizophrenic symptoms and signs appeared in conditions other than schizophrenia. Hence, in 1978, Hays could continue to theorise that several disease entities might be included in schizophrenia. And at the same time hypothesise that 'puerperal' schizophrenia was a separate disease (Hays, 1978). Yet much remained theoretical and speculative. Hence, Stephens noted:

> It is possible that the syndrome variously labelled reactive psychosis, nonprocess schizophrenia, schizophreniform psychosis, acute delusional psychosis and acute schizoaffective psychosis should be considered a separate illness not of the genus schizophrenia (1978, p. 41).

In the same vein, Fowler would suggest that the prognostically favourable 'remitting schizophrenia' was not a diagnostic entity. Rather, it was thought to be 'a heterogeneous mixture of mania, unipolar depression, and typical schizophrenia' (Fowler, 1978, p. 76).

Others would occasionally announce misdiagnoses of rare conditions such as metachromatic leukodystrophy (Anon, 1978b), or that distinguishing between brain-damaged patients and schizophrenic patients remained a significant clinical problem (Goldstein, 1978). And for some, the symptoms and signs in schizophrenia were not necessarily discretely different from functioning in normal people or from other nonschizophrenics (Strauss et al., 1974b). In 1989, Putnam would declare that the finding that multiple personality disorder patients had been misdiagnosed as suffering from schizophrenia had been replicated several times (Putnam, 1989).

To the 1930s' medical profession at least, there was nothing surprising about this kind of uncertainty. *The Lancet*, for instance, noted that 'the best psychiatric opinion' in all countries regarded schizophrenia as lacking unity. It was 'a provisional grouping, a congeries of biological types of morbid reaction' (Anon, 1933, p. 545). It was 'not a definite disease, and almost certainly includes heterogeneous conditions' (Anon, 1938, p. 1184). And for schizophrenia 'the diagnosis was apt to be wrong' (Anon, 1934, p. 985). The attempted separating out of multiple conditions, many with their own conceptual difficulties, simply took place within and in parallel to the conceptualisation of schizophrenia. However, as time passed, such problems could often be recast as a narrative that portrayed advances in knowledge. In 1977, for example, Howard Goldman could present a progressive narrative in which 'Toxia delira, general paresis, pellagra, and amphetamine psychosis are all forms of dementia praecox which have been distinguished from idiopathic schizophrenia' (1977, p. 3). Yet even progressivists such as Goldman continued to favour a nonunitary disease hypothesis. Goldman called for a re-examination of the entire phenomenology of schizophrenia, coupled with a search for previously 'unknown associated signs and symptoms' (ibid).

Other than briefly referencing many of these conditions, we cannot give them further attention here. But for many twentieth-century psychiatrists, at times schizophrenia simply was simply autism, epilepsy, encephalitis and the postpartum schizophrenic psychoses (Zilboorg, 1929). It was the so-called 'symptomatic schizophrenias' and 'paranoid schizophrenia of alcoholics' (Ellard, 1977, p.16).[9] Twentieth-century schizophrenia was not simply mistaken for multiple conditions, theoretical or otherwise. It often was them.

Conclusion

In 1911, Bleuler mentions in passing the use by a patient of an expression/word that Bleuler himself claims he cannot understand. The word cited is 'botanized' (Bleuler, 1911/1950, p. 155). The patient's voice is now lost to history. Yet the word remains symbolic. For in conceptualising schizophrenia, Psychiatry was inspired by botanical taxonomy as a way of ordering its observations. However, what worked reasonably well for slowly evolving plants was problematic in the domain of madness. And the process of classification encountered significant obstacles for those wishing to conceptualise schizophrenia.

As we have just seen, the presence of heterogeneous conditions in the same pathological space came into play. Conditions such as those variously referenced as influenza, epilepsy, autism, toxic infections, iatrogenic effects, encephalitis, manic depressive psychosis and postpartum psychosis all shared and criss-crossed the phenomenological space through which schizophrenia was mapped. Arguably, one historical explanation for all this confusion stands out from the rest; in 1938, Henry Ey could be found claiming that both schizophrenia and dementia praecox appeared to reference a plurality of states due to a plurality of aetiologies. Ey (1996) felt that psychiatry had been the victim of doctrinal dogmatism.

Compounding this problem, the phenomenology under investigation often fluctuated across time. No sooner had definite subtypes been observed on initial investigation, or during a period of stasis, when the phenomenology changed again. Twentieth-century psychiatry did not seek to abandon its subtypes. Each failure resulted in renewed attempts to break the concept down again. Loose and unclear conceptual boundaries inevitably led to inconsistency and ultimately communication problems.

With no fixed concept, regionalism and private classification inevitably characterised part of schizophrenia's conceptualisation. All of the above concerns meant that taxonomy was highly problematic for the concept of schizophrenia. Yet as we shall see later, as the twentieth century unfolded, an increasingly hegemonic North America would attempt to overcome such problems. With the landmark publication of the American Psychiatric Association's DSM-III, psychiatry would declare that its classification and conceptualisation of schizophrenia had become much more rigorous (American Psychiatric Association, 1980).

6

Myth and Forgetting: Bleuler's 'Four As'

In this chapter we now turn to examine how a process of historical forgetting and myth-making further compromised the conceptualisation of schizophrenia in the twentieth century. As previously noted, the Swiss psychiatrist Eugen Bleuler (1857–1939) first coined the term schizophrenia in 1908. In doing so he described one of its most important characteristics as a splitting, or tearing up, of the psychic functions. Bleuler subsequently outlined his concept in detail in his 1911 text *Dementia Praecox or the Group of Schizophrenias*, which is some 500 pages in length (Bleuler, 1908, 1911/1952). As this chapter now explores, recent generations of psychiatrists frequently describe and synopsise Bleuler's schizophrenia in terms of the 'four As' mnemonic. Namely: disturbances of affect, associations, ambivalence and autism. Therefore, it is not uncommon to find statements such as the following from an article entitled 'Notes on the History of Schizophrenia':

> It was Bleuler who first coined the divisive term 'schizophrenia' in 1911. Bleuler defined schizophrenia with his four 'A's', referring to the blunted Affect (diminished emotional response to stimuli); loosening of Associations (by which he meant a disordered pattern of thought, inferring a cognitive deficit), Ambivalence (an apparent inability to make decisions, again suggesting a deficit of the integration and processing of incident and retrieved information) and Autism (a loss of awareness of external events, and a preoccupation with the self and one's own thoughts) (Kyziridis, 2005, p. 45) (n.b. Kyziridis is in error about the date 1911).

Another example is the following:

> The 20th century ended without a resolution of the debate about the supremacy of Schneider's psychopathological conceptualisation of schizophrenia (the first-rank symptoms) over Bleuler's 'four As'(disorders of association and affect, ambivalence and autism) (Ceccherini-Nelli and Crow, 2003, p. 233).

And yet another reads:

> His fundamental, or basic symptoms, the four A's, i.e., loosening of associations, inappropriate affect, ambivalence, and autism were to become the most extensively employed diagnostic criteria of schizophrenia (Ban, 2004, p. 754).

However, this mnemonic, although useful in opening Bleuler's theory to a broader public, is somewhat inaccurate. And, as will be shown, it distorts Bleuler's thinking. Moreover, it does so in a way that is almost certainly symptomatic of a deeper failure to think about symptoms in a historically minded way. Bleuler never used the expression 'four As', nor did he consistently give priority or precedence to any four symptoms. As noted, for example, in 1908 Bleuler declared splitting or tearing of the psychic functions to be the outstanding symptom of the group of 'schizophrenias':

> *Ich glaube namliche das die Zerrei ung oder Spaltung der psychischen Funktionen ein hervorragendes Symptom der ganzen Gruppe sei* (I believe that the tearing up or the splitting of the psychic functions is the outstanding symptom of this whole group) (1908, p. 436).

By 1911 Bleuler dropped 'tearing up' but continued to stress the importance of splitting. Significant splitting was found in every case. However, Bleuler now declined to call splitting a symptom per se. Instead, he preferred in the discussion of schizophrenia's definition to call splitting one of its most important characteristics.

Hence, consider his opening comments in *Dementia Praecox or the Group of Schizophrenias*, 'Chapter 1—The Fundamental Symptoms':

> The fundamental symptoms consist of disturbances of association and affectivity, the predilection for fantasy as against reality, and the inclination to divorce oneself from reality (autism) (1911/1952, p. 14).

Here he stated three of the supposed four, omitting among numerous possible candidates ambivalence. The latter was attacked in the same year by Jung. He argued that it could not be put on the same level as the schizophrenic splitting of the psyche (Jung, 1911/1972, p. 198).

Subsequently, Bleuler listed the fundamental symptoms as belonging to two subsections: the simple functions and the compound functions. In the former, he listed three simple functions: association, affectivity, and ambivalence. This time, ambivalence was included but autism was missing. Once again, there was no fourth 'A'. Instead, in this case, autism was found in the compound functions subsection. Autism was given no exclusive priority over the other symptoms also found there. These pertained to and were classified under (b) attention, (c) will, (d) the person, (e) schizophrenic 'dementia', and (f) activity and behaviour. These symptoms ranked equally. Moreover, there was also an entire chapter on accessory symptoms, which derived from the fundamental symptoms and where, for example, in relation to autism, Bleuler noted that 'autistic thinking is directed by affective needs ...' (1911/1952, p. 67). Certainly, at a glance it is easy to get the impression that the symptoms appear to be listed in priority. This is an impression strengthened by the presence of two neologisms and the fact that these symptoms were not insignificant. However, as we shall see shortly, such a reading remains problematic.

For Bleuler, the fundamental symptoms were supposedly characteristic of schizophrenia, whereas the accessory symptoms may also appear in other illnesses. However, Bleuler later drew further alternative distinction between 'primary' and 'secondary' symptoms. Primary symptoms were theorised to stem directly from the disease process, whereas secondary ones were thought to occur when the sick psyche reacts to some internal or external process. And Bleuler, when outlining, in Chapter 10, his theory of symptoms, now seemingly changed his mind. He placed affectivity with a list of other symptoms in the secondary symptoms category. Disturbances of affect were no longer given precedence:

> Currently, I consider the disturbances of affect as secondary symptoms but in doing so I am well aware that I am in disagreement with the usual conception of schizophrenic deterioration (Bleuler, 1911/1952, p. 353)

Instead, he placed in the section on primary symptoms only disturbances of association. These were accompanied by a small discussion on the probability of other symptoms, such as 'clouded states' (*Benommenheitszustande*, very loosely a state of absence/taken-awayness), which he had earlier classified as accessory, and which Kraepelin had noted was present at the beginning of dementia praecox (1893, p. 439). Therefore, as the only symptom characterised as both fundamental and primary, disturbances of association were clearly, for Bleuler, the most important symptom of schizophrenia.

To complicate matters slightly and demonstrate this inconsistency further, it should be noted that in 1912 Bleuler wrote a paper on autistic thinking. He now appeared to upgrade its importance:

> One of the most important symptoms of schizophrenia is the preponderance of inner life with an active turning-away from the external world. The most severe cases withdraw completely and live in a dream world; the milder cases withdraw to a lesser degree. I call this symptom 'autism' (Bleuler, 1912/1951, p. 397).

And in 1926, in a short address to a conference in Switzerland, Bleuler does come close to delineating the four symptoms. Troubles of association and emotion, 'including autism', were declared as cardinal symptoms (Bleuler and Claude, 1926/2001). However, ambivalence was given as something often present rather than always present.

In any case, by 1930 Bleuler had once again referred to the secondary status of autism:

> According to our conception, we can distinguish in schizophrenia primary and secondary signs. Most of the symptoms described by Kraepelin, such as autism ... are secondary signs (1930b, p. 203).

Indeed, because people misunderstood what the name autism meant, he would also rename the symptom as dereism, considering it a thinking that disregards reality (Rapaport, 1951, p. 397). One of the so-called 'As' was now a 'D', with a slightly modified theoretical orientation. In the aforementioned 1930 paper, Bleuler declared, 'We consider as the main primary signs, both certain disorders in affectivity and in associations, which we have described upon other occasions' (1930b, p. 203). It should be stressed in passing that the concept of autism has no inherent historical unity. It has varied historically and has been subject to many interpretations since it was first formulated. Jung declared it the equivalent of the autoeroticism of Freud (1911/1972, p. 198). Minkowski in 1926 would split it into autisme pauvre and autisme riche (Bleuler and

Claude, 1926/2001). Bleuler himself derived it from Freud and Pierre Janet. He had initially called it alternatively ipsism and autism (Jung, 1974), although as the comment above notes, Kraepelin, too, played a role. Ultimately and perhaps facilitating its persistence somewhat, the term autism became vastly inflated. It encroached on faith, religion, the symbolic, the mythological and all that Bleuler deemed irrational or unobjective (Scharfetter, 2006).

Splitting

Moreover, as noted earlier, although it was no longer considered a symptom per se, splitting remained for Bleuler a very important characteristic of schizophrenia. If one reads the 1911 text closely, rather than over-relying on the index of symptoms, its presence is clearly stamped on the concept. This occurs even as relegated behind 'loosening of the associational structure':

> The splitting is the prerequisite condition of most of the complicated phenomena of the disease. It is the splitting which gives the peculiar stamp to the symptomatology. However, behind this systematic splitting into definite idea-complexes, we have found a previous primary loosening of the associational structure which can lead to an irregular fragmentation of such solidly established elements as concrete ideas. The term, schizophrenia, refers to both kinds of splitting which often fuse in their effects. (Bleuler, 1911/1952, p. 362)

Furthermore, in the theoretical section where Bleuler signalled association disturbance as the only primary symptom, his subsequent list of 'secondary symptoms' catches the eye. It now placed first in the alphabetically numerated list, before affectivity, autism and ambivalence, what the English translation misleadingly calls 'the train of thought-splitting' (Bleuler, 1911/1952, p. 355). This translation somewhat obscures the original German '*Gedankenablauf. Spaltung*' (Bleuler, 1911, p. 290). This translation should have read, 'The train of thought, splitting' or, more literally, 'The train of thought. Splitting', as it more accurately reflects the fact that both concepts (and indeed the above passage) are dealt with in this first section. Hence, one might want to justify a 'four As' mnemonic based on the structure of the theoretical section in conjunction with the structure found in Chapter 1. However, the mnemonic would ignore the importance Bleuler placed on splitting.

Therefore, although there are many symptoms beginning with 'A' in Bleuler's work, it is clear that the 'four As', as some kind of fundamental conception or abstraction of schizophrenia, cannot be seen as representative of Bleuler's writing. Not least because at most times his thought was in flux—as is the scientist's privilege. From his 1911 text in particular, it is unsatisfactory to derive the 'four As'. As such, Bleuler left somewhat unresolved an earlier complaint by Adolf Meyer in 1908. Meyer wrote that 'Bleuler does not define what his "primary symptoms" are, and it is therefore impossible to consider the justification of this logical manoeuvre' (1908, p. 247).

Bleuler Condensed

Nonetheless, after 1920 psychiatrists were already beginning to attempt to synopsise and condense Bleuler's thought and writing. Using the letter 'A' as a key, possibly because it grouped the two neologisms autism and ambivalence, seemed to be one such way of doing so. One example is Hans Prinzhorn's (1922/1971) *Artistry of the Mentally Ill*. In that work, approximately 75% of the pictures investigated originated from people classified as suffering from schizophrenia. Deliberately aiming at a wider public, Prinzhorn had summarised schizophrenia briefly, 'with special regard for psychiatrically unprepared readers' (1922/1971, p. 38). In a brief two-page overview, Prinzhorn gave autism as the primary symptom of schizophrenia, and 'affective ambivalence' as another. He additionally noted that 'it is impossible to establish emotional contact with a schizophrenic' (1922/1971, p. 39) and that 'inadequate emotional expression is commonplace ...' (ibid). Of affective ambivalence, he noted that 'emotional ambivalence is part of it' (ibid), and he further distinguished pure ambivalence itself:

> The same object is perceived and used in very different ways so that it remains logically incomprehensible why one conception does not necessarily exclude the other. Persons are also treated ambivalently (ibid).

Finally, he detailed 'associative loosening', although he added that it was probably only a derivative symptom due to the aforementioned basic traits. As with Bleuler's writing, it is impossible here to extract any definitive number of typical symptoms. Prinzhorn's summarised list of symptoms can be variously enumerated as three or four. However, it comes first to our attention in its condensing of Bleuler's 500 or so pages. And it is further notable for its alphabetic emphasis and for the discarding of all other symptoms in Bleuler's work.

None of this is to say that Prinzhorn's approach was the exclusive, or even the dominant, way in which Bleuler's symptoms were condensed. Later, in 1926, M. Wizel, Chief Psychiatric Medical Officer of the Hospital Cyste in Warsaw, in his article 'The Crude Forms of Schizophrenia', also attempted to describe Bleuler's cardinal symptoms. These he identified as loss of contact with reality and autistic thinking (which he saw as different but intimately related), followed by dissociation of thought and, finally, 'affective disorders and ambivalence' (*les troubles affectifs, et l'ambivalence*; Wizel, 1926, p. 446). Wizel, writing in French, did not mention the 'four As'. This was an impossibility given that he spoke not of 'loosening of association' (*associations relaches*), as we might expect (see below; Bourgeois, 1999), but of 'dissociation of thought' (*dissociation de la pense*). This, of course, lent itself more to Bleuler's emphasis on splitting. Indeed, he shored up this latter sense with the presentation of a case study of dual personality, which he saw as a very marked form of this process.

Furthermore, Wizel stated that these symptoms were only *among* the most important (*parmi les premiers*), not that they were the exclusive cardinal symptoms. One thing driving Wizel's condensation, it seems, is that Wizel was attempting to define from four case studies—in a short article—the 'crude forms' (*formes frustes*) of schizophrenia. These supposedly lay conceptually somewhere between the schizoid and schizophrenia. As such, the crude forms were considered less overt than schizophrenia. And because the patients were considered lucid they naturally had to have fewer or different symptoms. Wizel, as such, excluded most symptoms such as hallucinations and delusions. Unsurprisingly, *benommenheitszustande*, which alluded to a state of mind that did not fit in with the idea of lucidity, also could find no place in Wizel's (1926) system. The latter omission occurred, even though *bennomenheitszustande* had been recognised elsewhere by authorities such as Régis (1914).

We find another alternative contraction in the work of Silvano Arieti, one of the most encyclopaedic psychiatrists to have written on schizophrenia in the 1950s. Arieti's 1955 edition of *Interpretation of Schizophrenia* carried a declaration concerning Bleuler's contribution to psychiatry:

> The most important contributions of Bleuler were those related to his study of the process of association and disturbances of the affective life, the concepts of autism and ambivalence, and his interpretation of negativism (1955, p. 15).

For Arieti, these symptoms were not just a summary. They were a subjective listing of Bleuler's most important contributions to schizophrenia research, which Arieti saw as an extension of Kraepelin's work on dementia praecox. From such an authority, this assessment would have carried some weight. Yet, despite this progression in the importance attached to such symptoms qua contributions to research theory, Arieti had additionally now mentioned negativism. And nowhere did he mention the shibboleth 'four As'. Elsewhere, a close reading of Bleuler in 1951 by David Rapaport interpreted autism along with attention and other symptoms as complex functions. Rapaport declared with emphasis that,

> Though his formulations are somewhat equivocal, Bleuler believed that the basic symptom of schizophrenia is loosening of associations based on an organic process, and is the prerequisite of the other basic and the accessory symptoms (1951, p. 582).

And, indeed, in the same year, a review of Bleuler's newly translated 1911 text in the *Journal of Consulting Psychology* stated that for Bleuler the primary phenomenon of schizophrenia was a disturbance of associations, a 'loosening' of the thinking processes. All other symptoms were regarded as secondary (Anon, 1951).

This lack of agreement is not entirely surprising. Indeed, explicit mention of the 'four As' appears absent from most of the early twentieth-century writing on schizophrenia in English. For example, for all the literature on schizophrenia written in the *American Journal of Psychiatry* with schizophrenia in the title (prior to 1960), the 'four As' simply do not seem to exist. This negative finding extends to a large number of other printed materials that are too numerous to mention. Nor do we find such a mnemonic in either of two synopses aimed at Bleuler's work (Ey, 1926/1969; Kline, 1952). Even though the latter, *Synopsis of Eugen Bleuler's Dementia Praecox or the Group of Schizophrenias*, by Nathan S. Kline was aimed at 'time hungry souls (medical students, candidates for the American Boards, etc.) who can't squeeze in the real thing at just this juncture ...' (1952, p. i).

The Rise of the Mnemonic

What appears to be the first reference to the 'four As' occurs when Charles Hofling and Madeleine Leininger mention it in their 1960 book *Basic Psychiatric Concepts in Nursing*:

> In considering some of the common characteristics of schizophrenia, one may organize the material around a convenient memory device,

> known as 'Bleuler's Four A's'. These characteristics, to be defined in the course of the discussion, are: apathy, associative looseness, autistic thinking, and ambivalence. For the student's convenience, a fifth item, auditory hallucinations may be added to the list, although it is actually a sharply defined symptom rather than a 'characteristic', and it is less basic and less nearly universal than the others (1960, p. 298).

This reference is notable because for the first time it appears that we have a reference to the term 'Bleuler's four As'. The 'four As'—the same edition noted—were taken after 'the great Swiss psychiatrist, Eugen Bleuler (1857–1939), who first described them in combination' (ibid). Furthermore, the text suggests that it was a useful mnemonic device for organising some common characteristics of schizophrenia. Hofling and Leininger also introduce auditory hallucinations as a fifth possible 'A'. But they refrain from explicitly declaring the 'five As' or from calling auditory hallucinations characteristic. In doing so, the authors differ from the aforementioned work by Arieti, who had included as a fifth characteristic, negativism. Moreover, they focus on 'inappropriate emotions' of the schizophrenics—'emotional (affective) responses' (ibid). Furthermore, and as articulated in later editions, they emphasised apathy, whereby feeling manifested by the patient seems to be out of keeping with the ideas being expressed, or the amount of emotion shown is unusual. As we shall see, it was from this time on, and only a decade after the first English translation of Bleuler's 1911 textbook, that the literature now coalesced somewhat around the idea of the 'four As'.

By 1966, Arnold H. Buss, in *Psychopathology*, would also claim that Bleuler believed 'that there were four fundamental symptoms: association, affect, autism, and ambivalence' (1966, p. 187). But Buss adds that no two authorities listed identical symptoms as being fundamental. Buss further adds that, with respect to the aforementioned list of symptoms, 'The last two are rejected as being fundamental by most authorities, and there is some doubt about the first two' (ibid).

Buss did not speak of the 'four As'. Instead, deferring to Mayer-Gross et al. (1955), he thought an attempt to find one fundamental psychological disturbance underlying all symptoms was asking too much. This was all the more so if there was doubt whether the 'present concept' of schizophrenia did not comprise of several diseases. Clearly, though, consensus on what Bleuler historically considered important was now appearing to gravitate towards four symptoms. In 1972, Theodore Sarbin published a critique of the entire concept: 'Schizophrenia is a Myth, Born of Metaphor, Meaningless'. In doing so, he cited and repeated Buss's critique and further stated:

> Bleuler named the disorder, and defined it in terms of The Four A's: weak associations, inappropriate affect, ambivalence, and autism (preoccupation with oneself). The Four As continue to underlie modern definitions (Sarbin, 1972, p. 21).

Hence, just as the 'four As' were coming into being—as something Bleuler defined—they were beginning to be consigned to history.

Four years later, in the *American Journal of Nursing,* Helen M. Arnold (1976) explicitly cited the third edition of *Basic Psychiatric Concepts in Nursing* (Keyes and Hofling, 1974) in an article titled 'Working With Schizophrenic Patients. Four A's: A Guide to One-to-One Relationships'. Arnold observed:

> Most nurses are taught Bleuler's classic 'Four A's' to describe the characteristics of schizophrenia: inappropriate affect, loose association, ambivalence, and autism (1976, p. 941).

In doing so, Arnold removed the former term 'apathy' and restores the term 'affect'. Moreover, she affirmed that although the 'four As' were useful as a mnemonic device, they also conveyed a negative attitude by focusing on pathology. Hence, citing critiques of schizophrenia by Scheff (1966) and Sarbin (1972), Arnold further noted problems with the legitimacy of each of the As in question. And instead suggested that 'it would be more helpful for patients and nurses to focus on 'Four A's' which are therapeutic—acceptance, awareness, acknowledgment and authenticity' (Arnold, 1976, p. 941). Nevertheless, what Keyes and Hofling described as 'some common characteristics' Arnold now called classic characteristics. By 1986, Thomas Oltmanns, John Neale, and Gerald Davidson could be found observing that,

> Bleuler argued that there were four primary symptoms of schizophrenia: loosening of associations (disorganized speech), blunted or inappropriate affect, ambivalence (the simultaneous expression of opposite emotions, attitudes, or wishes toward a given person or object), and autism (a preference for fantasy over reality). These came to be known as the four As (1986, p. 233).

Apart from adding new erroneous distortions, such as declaring the symptoms to be primary and characterising loosening of associations as disorganised speech, the authors now added the following critique to the fictional synopsis of Bleuler's work:

> Unfortunately, they are all somewhat difficult to identify with an acceptable level of reliability. It is very difficult, for example, to decide whether a patient's mood is blunted, slightly depressed, or contemplative. In fact, the patient may be either preoccupied. [Or bored??] Because of the ambiguity surrounding these kinds of criteria, schizophrenia came to be a very broad, poorly defined category. DSM-III has reversed that trend (ibid).

Earlier, Robert Kendell, implicitly referencing the 'four As', had similarly declared that Bleuler's fundamental symptoms were 'peculiarly intangible and difficult to define'. Hence, 'for this reason none can be a welcome component of any operational definition' (Kendell, 1972/1975, p. 14). Similarly, in 1980, Heinz E. Lehmann remarked that the 'four As' were not particularly helpful in making a clinical diagnosis of schizophrenia. This was 'because they are so general and, in the case of autism and ambivalence, refer more to existential attitudes than to clinical symptoms' (Lehmann, 1980a, p. 1153). As such, by this time the broadness and diagnostic unreliability of Bleuler's concept of schizophrenia was now being attributed to four primary symptoms whose existence was difficult to identify. Yet these actually only caricatured Bleuler's work. In fact, as we shall see, Bleuler would never have relied on these four symptoms for diagnosis. However, the proposed idea of dispensing with Bleuler's formulation now presented a self-deceiving illusion of progress and remedy.

The Mnemonic Contested

The incorrectness of the 'four As' did not go entirely unnoticed. It was suspected by Andreason et al. (1984), who, seemingly divorced from primary sources, thought it probably incorrect. Several generations of psychologists and psychiatrists, they thought, had fallen under the mnemonic's spell:

> a somewhat loose reading of Bleuler in the United States has given us the Bleulerian 'Four A's'. ... Based on a widely accepted (and probably incorrect) understanding of Bleuler, the last several generations

> of psychologists and psychiatrists have been taught to recognize schizophrenia in terms of the Four A's and to see formal thought disorder as the most important among the fundamental symptoms (Andreason et al., 1984, p. 199).

In a similar vein, Shean (2004) spoke of three simple functions plus one compound function (autism), while simultaneously claiming Bleuler had downgraded autism by 1923. In doing so, Shean was of the opinion that Bleuler identified between four and six fundamental symptoms, 'depending on how one interprets his writings' (2004, p. 15). Other authors, also sensing the problem, have tried to qualify the mnemonic, or have even given primacy to one among the four. For example, Martin Harrow and Donald M. Quinlan, in *Disordered Thinking and Schizophrenic Psychopathology*, noted of Bleuler:

> In his view, the major symptoms of schizophrenia were the 'four As' affect, association, autism, and ambivalence. While this included affective, interpersonal and other dimensions of the schizophrenic's behavior, his central emphasis was on associations (1985, p. 7).

Similarly, Carpenter and Buchanan declared that,

> Bleuler introduced the concept of primary and secondary schizophrenic symptoms: his four primary symptoms (the Four As) were abnormal associations, autistic behavior and thinking, abnormal affect, and ambivalence. Of these four symptoms Bleuler viewed as central to the illness the loss of association between thought processes and among thought, emotion, and behavior (1995, p. 889).

But such suspicions and attempts to resolve the inconsistency have been to no avail, as yet another example shows:

> Main symptoms were the loosening of associations, disturbances of affectivity, ambivalence and autism. Accessory symptoms were, for example, delusions, hallucinations, alteration of personality, language and writing as well as catatonic symptoms. In Bleuler's interpretation the four main symptoms ('the four a's') were 'exaggerations of physiological phenomena' (Stotz-Ingenlath, 2000, p. 157).

And as shown above (Ban, 2004; Ceccherini-Nelli and Crow, 2003; Kyziridis, 2005), the error remained widespread, in some cases even masquerading as history. As noted earlier in the work of Sarbin (1972), it is also clear that the error is not peculiar to orthodox psychiatry. Critiques of schizophrenia, for example Richard P. Bentall's *Madness Explained* also reiterate it:

> In an attempt to characterize this unity, he identified four subtle symptoms which he believed to be fundamental to the illness, and which have since been known to the English-speaking psychiatrists as Bleuler's four 'As' (2003, p. 23).

Nor was the error unique to North American psychiatry. In French psychiatric literature, for example, Bourgeois noted 'that E. Bleuler and the Anglo-Saxons call the 4 A: Autism, blunted Affects, loosened Associations, Ambivalence (*ce que E. Bleuler et les Anglo-Saxons appellent les 4 A: Autisme, Affects emousses, Associations relachees, Ambivalence*)' (1999, p. 14). And, to add a twist, in contemporary German psychiatric literature, Scharfetter, in *Eugen Bleuler: Leben und Werk*, discusses not four but three As. Hence, 'the 3 A's: disorder of association, disorder of affect, autism (*die 3A: Assoziationsstorung, Affektstorung, Autismus*)' (Hell et al., 2001, p. 34). All this, in turn, fed into the way people theorised about schizophrenia and, as we have seen with Arnold (1976), its treatment. For example, to compound this historical error, further parallels have been extended and drawn by one author to negative symptoms, which are purportedly (but most definitely not) synonymous with the 'four As': 'the negative symptoms (synonymous with Bleuler's Four As)—including anhedonia, amotivation and affective blunting' (Singh, 2005, p. 413).

Ahistoricism in Schizophrenia Research

Perhaps the mnemonic predates its apparent first occurrence in 1960. It is possible that the 'four As' might have been in use within the psychiatric community before this time. However, at present, such evidence appears absent. In any case, such a caveat does not undermine the core validity of our findings. The concept of the 'four As' has no basis in the historical texts it purports to emerge from. And even if it had, the representation of the 'four As' has not remained constant. There have been at least five (apathy is not the same as disturbances of affect) and as few as three. Varying and erroneous simplifications for the same A have occurred. Disorganised speech, for example, is not the same as loosening of associations; Bleuler considered abnormalities of speech and writing to be separate symptoms. In this process, the 'four As' have moved chronologically from being important contributions to something defined by Bleuler. Consequently, Bleuler's work has been judged because of a mnemonic, which he appears to have played no part in creating. The simple dismissal of Bleuler's 'four As' in theorising was not only an error, but also detracted from analysis of whether Bleuler's group of schizophrenias itself was ever justified. Simultaneously, belief in the 'four As' has probably played a role in some of Bleuler's other theorised primary symptoms, such as *Benommenheitszustande* (almost unheard of in contemporary literature) being forgotten. Similarly, other important phenomena such as the splitting of personality were also neglected, ignored, sidelined, or forgotten.

Explaining the 'Four As'

A mnemonic may, of course, come into being at any time. Yet, it remains interesting to ask why it was that the 'four As' appeared to have emerged in writing only in the second half of the twentieth century. Let us consider not just why the 'four As' seemed to come to prominence at this time, but also what function they might serve in persisting.

First, let us remind ourselves by way of context that by the time of the mnemonic's emergence, a decline in Bleuler's orthodox conception of schizophrenia and its influence had been ongoing for some time. In fact, the continuous dilution, merging and transformation of his ideas had occurred from as early as 1911. This had occurred under the influence of theorists such as Adolf Meyer, Sigmund Freud, Carl Jung, Harry Stack Sullivan and even Emil Kraepelin. Hence, although many psychiatrists were ostensibly diagnosing 'Bleuler's schizophrenia' (not everyone—some continued to prefer dementia praecox as a diagnosis), they were, in fact, drawing from a variety of theoretical sources. There were numerous influences in making their diagnosis. The terminology of competing concepts, such as Kraepelin's dementia praecox or Meyer's paragasia, had failed to win enough support to defeat Bleuler's terminology. Yet the theories behind them retained considerable influence. Indeed, in 1927 Minkowski would even complain that the term schizophrenia was often, and wrongly, confused with psychoanalysis (Minkowski, 1927). This meant that schizophrenia in pre-1950s psychiatry was not quite as Bleulerian in practice as might be supposed. This applied even to Eastern Switzerland. There, Aubrey Lewis expressed surprise in 1938 that Bleuler's 'psychopathological' approach had little influence (Angel et al., 2003, p. 94).

Furthermore, with growing Anglo-Saxon hegemony, later generations of English-speaking psychiatrists, unlike earlier generations, were, for various reasons, increasingly less likely to have the language skills to read the information contained in the many untranslated yet still important German texts.[1] The link between the psychiatrist and the primary source was slowly being severed. This problem held true even if the work in question was a key text. In 1952, psychiatrist Nolan C. Lewis, for example, had noted the existence of digests for the many students unable to read German at a time when Bleuler's key text, *Dementia Praecox or the Group of Schizophrenias*, had no English translation (Bleuler, 1911/1952, p. i). Compounding this problem, as observed by Joseph Zinkin in the translator's preface justifying the translation, was the fact that most North American libraries did not even contain a copy of Bleuler's 1911 work. At most, libraries seem to have contained only the incomplete description of schizophrenia that is present in the 1924 *Textbook of Psychiatry* (Bleuler, 1916/1924). Zinkin further noted that everybody assumed that the 1911 text had been translated and that everyone else had read it.

All this, among other things, facilitated a climate of diagnostic confusion (Campbell, 1929/1934; English, 1934), which ought to have thrown the validity and reliability of the concept into stark focus. In fact, these contradictions and tensions would be swept under the carpet of therapeutic optimism and lie unexposed until after 1950. For whatever the validity of schizophrenia, whatever its reliability, the triad of therapeutic approaches that entered late-1930s psychiatry (i.e. electroshock,[2] insulin coma treatment and lobotomy) had temporarily reduced the perceived need to understand or justify schizophrenia in favour of 'treating it'.

With the translation of Bleuler's canonical text, English readers were now for the first time able to read Bleuler in translation. They could interpret and study the historic work closely for themselves and reduce their dependence on the interpretations of other authors. Nevertheless, professionalism continued to exert itself on psychiatry and other related disciplines. Hence, 'time-hungry' students, who also had to deal with an ever-increasing number of declared psychiatric disorders, continued to read summaries for exams, which became an increasingly more important part of their training. Therefore, in addition to the emergence of new reinterpretations of this rich new source of descriptive psychopathology, digests for Bleuler's work remained in demand. So, even when the 1911 text was translated for the first time in 1950, it was quickly followed (as noted above) by two brief synopses by Kline and Ey. The result of this need to abbreviate, condense, and abstract was twofold. First, it allowed errors to creep into psychiatric thinking on schizophrenia, and, second, it reduced the possibility for the discipline's critical reflexivity on these same errors. All this primed the second half of the century for the possible emergence of a new summarisation or interpretation of Bleuler's work such as the 'four As'.

Diagnosis and the 'Four As'

More recently, we have witnessed a growing commentary on the failure of the 'four As' in diagnosis. Bleuler's 'four As', as opposed to other problems with the concept of schizophrenia, were held to be of little use in facilitating diagnosis (Lehmann, 1980a, p. 1153). Similarly, the idea has been expressed that Karl Schneider's system of operational definition had been in battle with the 'four As' (Ceccherini-Nelli and Crow, 2003). Certainly, Bleuler's 'four As' have at times served as a useful rhetorical contrast to Schneider's first- and second-rank symptoms (again facilitating their perpetuation). However, it should be stressed that Bleuler had never said that all patients could be easily diagnosed. He had, in fact, argued that diagnosis was at times impossible (Bleuler, 1911/1952).

In his discussion of cases of 'simple schizophrenia', for example, Bleuler noted, 'There is no doubt that many simple schizophrenics are at large whose symptoms are not sufficiently pronounced to permit the recognition of mental disorder' (1911/1952, p. 238). Instead, he cautioned that this group was rarely found within hospitals and that in any case it might take days or years for suspicions to be confirmed. The confusion of adherents to the idea that the 'four As' were problematic in diagnosis may arise here, however, because Bleuler did note that the group 'possesses minor theoretical value inasmuch as it demonstrates the difference between the essential and accessory symptoms: the latter are absent in simple schizophrenia' (1911/1952, p. 236). But this comment is, in fact, inconclusive. For, once again, Bleuler does not name the essential symptoms that many would later attribute to the 'four As'.

Moreover, Bleuler was essentially arguing that such cases were rare, occurred outside the clinic, and that, in fact, hospital diagnosis should usually refer to paranoid, catatonic, and hebephrenic cases. For these cases, the array of symptoms, which he detailed elsewhere, could and should be used in their diagnosis. Latent schizophrenics, for example, were identified by the fortuitous observation of a delusion or hallucination (Bleuler and Claude, 1926/2001). Early psychiatrists, who diagnosed schizophrenia, seem to have understood this enough not to show overt preference for any 'four As' schemata.

If a problem of diagnostic confusion occurred in hospitals (and it did), it was not due to the 'four As'. There is little evidence of their preferential use as a tool of psychiatric diagnosis. It was instead a problem of diagnosis consequent on all the declared constellations of signs and symptoms, which Bleuler and various psychiatrists believed to be significant. Hinting that problems of diagnosis are found with the 'four As' carries an implicit implication that historical confusion in schizophrenia diagnosis can be sourced to the 'four As'. The 'four As' were consequently presented as a convenient but inaccurate scapegoat for past problems in diagnosis. The ease with which such comments are made, and accepted, perhaps also ensured the continued survival of the mnemonic.

Conclusions

Bleuler's symptom profile is too complex to outline here. And only a thorough and repeated reading of Bleuler's work will yield an understanding of it. Nevertheless, the 'four As' mnemonic was clearly a distortion and simplification of Bleuler's complex concept of schizophrenia. It seems then that through the naive use and abuse of the 'four As', knowledge had been displaced in psychiatric and allied cultures. It had been displaced in favour of a mnemonic device, which purported to be fact. Nevertheless, the mnemonic was functionally useful for those who applied it in a number of different ways. These included the naive display and transmission of assumed 'knowledge' to other disciplines such as nursing. Also included was the validation of theoretical preferences, by false analogy and so forth.

Worst of all, the 'four As' gave the illusion that Bleuler's thought was easy to interpret and without ambiguity and tensions. In fact, Bleuler's thinking contains many tensions. These can be traced, for example, to influential thinkers such as Kraepelin, Freud, and Pierre Janet, and indeed through to numerous other stimulating ideas, for example Wernicke's concept of sejunktion and the associationist psychology of Herbart (Scharfetter, 2006). The real strengths and weakness of Bleuler's work, and his interpretation of other thinkers' concepts, therefore seldom came up for consideration. Allowed to stand without critical re-evaluation, such a representation of Bleuler's schizophrenia implicitly undermined the validation of the concept itself.

This exposition of Bleuler's 'four As' is important because it serves as a useful historical example to illustrate one way in which psychiatry's understanding and representation of its own arch disorder had drifted free from the text that brought it into being. Another notable example would be the frequent conflation of the concepts of schizophrenia and dementia praecox, which, however similar, are not the same. Such 'synonymising' occurs despite attempts to signal important differences between Kraepelin and Bleuler (cf. Scharfetter, 1999, p. 32) and despite an early and accurate recognition that Bleuler's concept was broader (Hoch, 1912), irrespective of what Bleuler claimed. Such examples are important because there is no reason to think that the problem had not occurred right across an array of symptoms, features, and theories of schizophrenia (particularly those symptoms now more fashionable than the 'four As').

The ahistorical nature of psychiatric thought, however, was not something that twentieth-century schizophrenia researchers ever felt an urgent need to address. And journals as much as individual researchers seldom recognised the importance of simple historical principles such as using primary sources over secondary commentaries. Indeed, in an atmosphere where naive realism reigned, schizophrenia's history was frequently reduced to progressive narratives and the celebration of heroic psychiatrists. Understandably then, at times schizophrenia conceptualisation inevitably suffered. Schizophrenia became schizophrenia, myth, forgetting and all.

7
Social Prejudice

For much of the twentieth century there existed a general amorphous quality in the concept of schizophrenia. At various times, and in various places, this amorphous nature readily allowed schizophrenia conceptualisation to work in ways that seem antithetical to science. In various ways, schizophrenia conceptualisation permeated, and could be permeated by numerous cultural beliefs, values, imperatives, and attitudes that now disturb us. For to further their descriptions of madness, many in psychiatry and related professions, readily adopted society's prejudices as ancillary symptoms or signs. And over time a slow accretion of miscreants took place within the literature.

As we shall see, schizophrenia conceptualisation facilitated the pathologisation of sexual behaviour, race, and the family. It did the same for various other political and social deviances. And in one way or another, such people were all at odds with society. And in many senses, they all fitted within one archetypal category that Manfred Sakel recognised and used mid-century: moral insanity (Sakel, 1958). Or, as Bleuler put it, people 'who make our world uncertain' (1911/1952, p. 236).

Such flexibility is unsurprising. It is well recognised, for example, that a diagnosis of dementia praecox with Kraepelin could serve dual medical and administrative purposes (Noll, 2011). Moreover, psychiatric knowledge, to paraphrase Basaglia, is rooted in the prevailing moral order. It is, according to Dowbiggin (1997), almost invariably a blend of cultural attitudes, social values, political beliefs, and professional imperatives. Nevertheless, any full appreciation of schizophrenia and its conceptualisation is incomplete without documenting this side of the concept. For, at the very least, it contextualises the increasingly contentious status of the concept in the twentieth century. We need to detail it rather than merely acknowledge it.

Hard boiled

In the early twentieth century, society's institutional faith in the psychiatrist's ability to classify deviance—most emphatically in the lower classes—was expressed most clearly by one Herbart Harley. He did so in the *Journal of the American Institute of Criminal Law and Criminology*—seemingly after interacting with psychologists, psychiatrists, and their literature. One of the most common attributes of the dementia praecox case, argued Harley, was the inability to appreciate *moral* distinctions. Dementia praecox, he argued, produces burglars, automobile thieves, pick-pockets, counterfeiters, and yeggmen (safe-crackers). They were the hard-boiled guys of the criminal gangs. They banded together and made a profitable business of crime. From this class came the dementia praecox cases with sex complexes and those with paranoid tendencies gradually hardening into murderous intent.

Such criminals were finally brought in after committing a crass crime, usually murder in connection with robbery, or murder in connection with rape. The police, after many a terrible lesson, had learned that the dementia praecox case—of low intelligence—was extremely dangerous. Such cases shot on slight provocation. Most of the killings of officers, thought Harley, were of this type. So the police were becoming alert to picking up all sorts of eccentrics. The crank and the quarrelsome person once looked upon as *harmless* and *mirth-provoking*, were now under suspicion (Harley, 1921).

Harley appears somewhat less than liberal. However, it should be noted that he was, in fact, arguing in the paper for segregating those with dementia praecox, rather than upholding a then popular argument for *hanging* them—which he thought did little good. Indeed, his description of dementia praecox does not appear atypical, when contrasted with media stories around this time. For Harley, it was a *fact* that these dangerous persons were readily *diagnosable* at an early stage. He believed 'laboratory records' *proved* this incontestably. One could pick out a few cards from the laboratory records and say with reasonable certainty that one from that group will *murder* without provocation within six months. Two would murder within 12 months; three within 18 months, and so on. The prevention of crime was possible as one could forecast it from psychological and psychiatric tests (ibid). With such a belief, a diagnosis such as 'pfropfhebephrenie' could then be viewed as a fatal diagnosis indicating criminal propensity: 'One of the first great forward steps would be to prevent the victim of pfropfhebephrenia from getting out of Pontiac. No such case can safely be released if more than fourteen years of age' (Harley, 1921, p. 523). All this emerged out of a period when society was moving from the idea of treating delinquents as a single large class (Noll, 2011).

There was no shortage of authoritative reports to go on. In 1931, Karl Birnbaum could detail two types of 'schizophrenic': 'the passive asocial' and the 'active criminal' (*Der passiv-unsoziale und aktiv kriminelle Schizophrene*). These existed alongside the schizoid criminal (Birnbaum, 1931). For *The Lancet's* Clifford Allen (1936), false confession of murder indicated future propensity to murder, and was an early symptom of schizophrenia. While Gillies, in the *British Journal of Psychiatry*, would conclude that 'matricide is *the* schizophrenic crime' (Gillies, cited in Chiswick, 1981, p. 1279, my emphasis). The study involved just four matricides.

As Goldstein has observed, the power to classify is one of the most basic and even primordial of social powers. It is one claimed by and granted to the psychiatrist (Goldstein, 1990). And for much of the twentieth century, the public imagination, like Harley's, also held psychiatry as a force to classify and manage those presumed violently insane. This attitude seemingly then further tolerated and perhaps even fostered a less than imaginary violence. Such a disposition would certainly partially explain, alongside the need for spectacle, the front page of a *Washington Post's* nonchalant, and at best indifferent, 1938 account of 'diabolical madmen' being administered shock treatments (Gross, 1938).

Frequently, for the behavioural professions themselves, moral prejudice was subtly packaged within the descriptions of a subtype. This occurred in conjunction with descriptions of cases, signs, and symptoms; at times with a debatable ambiguousness—but more often not. In the *Psychology of Abnormal People*, John B. Morgan of Northwestern University would outline the 'simple schizophrenic reactions' and the 'hebephrenic reactions'—declaring both regression disorders. If the simple schizophrenic was born into a family with sufficient family support, they become the typical 'lounge lizard' or idle 'old maid' (Morgan, 1928). Otherwise, they became 'happy hooligans', hoboes, prostitutes, pseudogeniuses, cranks, and eccentrics (Morgan, 1928, p. 522). [Although later American Psychiatric Association (APA) president Theodore Blau would declare that 'The difference between schizophrenia and eccentricity may be the degree to which society approves or admires the observed behaviour' (Anon, 1970b, p. 167).]

More explicitly, Gregory Zilboorg's *ambulatory schizophrenics* presented a number of 'deficiencies'. For instance: hypochondriacal complaints, inefficiency, and tenacity in their inability to be productive and independent. For Zilboorg, cases of ambulatory schizophrenia were not infrequently sexual perverts, exhibitionists, transvestites, or fetishists. They were also considered by both 'the laity' and the medical profession as merely weak people, 'poor personalities', or 'psychopathic personalities'. Finally, he also considered them to be, at times, criminals. And at that mostly murderers: 'The nineteen- year old Volkman [executed at Sing-Sing], murderer of the little girl whom he first raped, comes to mind' (Zilboorg, 1941, p. 154). In doing so, an autonomous part of the personality seemed to act independently of the rest of the personality functions.

Alongside his work on insulin coma treatment, Manfred Sakel wrapped a moral reading of madness within in his own schizophrenia classification. He did so with the subtype moral insanity (Sakel, 1958, p. 35). In *Schizophrenia*, Sakel noted:

> We include moral insanity as a manifestation of the schizophrenic disease process because the moral precepts of man are an inseparable and integral part factor of the total personality and this can be stricken by disease just as easily as the part factors connected with the emotions or the reason (1958, p. 38).

Sakel was not being idiosyncratic. Bleuler had earlier subsumed and reconfigured the concept of 'moral insanity' into his 1911 formulations—as others such as Piquemal (1912) immediately noted—and journals still devoted space to it long after. For example, the *Revue Française de Psychanalyse* was still discussing its potential to cause *schizophrénie tardive* in the mid-30s (Pichon, 1936). Similarly, in 1890, Kahlbaum had suggested the term *Heboidophrenie* be applied to cases with less cognitive impairment and behaviour that is more antisocial (Berrios et al., 2003). But Sakel's text took things to its logical conclusion. It now made moral insanity an official subtype in its own right. Anyone diagnosed with schizophrenia, who stepped outside the moral worldview of the psychiatrist could now have their own subtype. As it happens, Sakel's star eventually faded. Insulin treatment for schizophrenia gave way to neuroleptics. Insulin treatment was ineffective, sometimes fatal, and among other things was linked to fetal disorders (Wickes, 1954). The subtype never found lasting recognition. Instead, *moral insanity* was reduced to a fossilised symbol in the research literature for all that had been possible in judging a patient's moral precepts. Nevertheless, such psychiatric judgements probably retained an implicit legacy for quite some time. Even the humanist psychologist Carl Rogers apparently admitted to R.D. Laing to having described schizophrenics as 'the most evil people in the world' (Burston, 1996).

Masturbation

As the concepts of katatonia, dementia praecox, and schizophrenia slowly unfurled, the social expediencies offered by classification had also readily intersected with prevailing judgements concerning sexual behaviour. Masturbation does not make the condensed symptom list provided by Kahlbaum in his 1874 definition of katatonia. Nevertheless, for Kahlbaum, the subject still occupies an important place in his monograph.

In particular, Kahlbaum deemed sexual overstimulation and a history of early masturbation in men as worthy of aetiological inference. Hence, there are numerous references to it. Patient Benjamin L., for example, who presented little in the way of aetiological clues, had weakened his nervous system through masturbation (Kahlbaum, 1874/1973, p. 7). Kahlbaum also duly noted that Adolph L. had masturbated at puberty for three years (Kahlbaum, 1874/1973, p. 18) and that Paul M. had indulged in masturbation since the age of 14 years (Kahlbaum, 1874/1973, p. 32). Elsewhere, like Paul M., Julius M. was diagnosed with early and long-continued masturbation. Kahlbaum, possibly influenced by general thinking with regard to sexual activity and general paralysis of the insane, accordingly saw masturbation as symptomatic.

Doubtless it was for such reasons that when the first dissertation on katatonia supported by Kahlbaum was written up, it too singled out masturbation in its patient history (Rust, 1879). Beyond the aetiological, Kahlbaum further claimed that masturbation augured an unfavourable prognosis. It did so because it represented an underlying pathological aggravation, within the sexual organs. Treatment for masturbation would include changes to diet, drugs, surveillance, moral hectoring, and catheterisation.

By 1883, Hammond would sensationally report that his patient could masturbate while in a state of stupor (Hammond, 1883). And a little later, Spitzka (1887) noted the general opinion among alienists that katatonia was one of the forms of insanity most frequently found in masturbators. In 1888, Ludwig Wille, in Basel, would also draw attention to Kahlbaum's *emphasis* of the role of masturbation in katatonia (Wille, 1888). In 1899, Lewis could reject the idea of epileptic katatonia. However, he still noted that the symptoms in question were 'closely associated with the vice of onanism' (Lewis, 1899, p. 238). And in 1902, Robert Jones would apparently assert that masturbation aggravated the condition (Anon, 1902).

Kraepelin pathologised masturbation and suspected a linkage with dementia praecox, but had failed to confirm a causal correlation in 1896 (Hall, 2003). Nevertheless, for Kraepelin, as Hare noted in 1962, masturbation still functioned as an indicator of the presence of dementia praecox in adolescents, prior to its own conceptualisation (Hare, 1962). For example, discussing the type of insanity supposedly brought about by onanism, Kraepelin noted that 'we see there without any difficulty the picture of dementia praecox' (1896, p. 51). In 1911, Bleuler, too, fails to find an association. However, he repeats the observation made by Schuele [sic] in 1901 relating cataleptiform attacks such as mutism to masturbation (Bleuler, 1911/1952). Possibly, this lingering inconsistency was exacerbated by extreme events in asylums of the period. On 17 April 1907, for example, in a letter discussing a possible catatonic case, Jung informed Freud that there were cases that die of autoeroticism. And that Jung (1974) had recently witnessed one such case.

Irrespective of the attitudes of Kraepelin and Bleuler's towards the idea, it remained in the literature. Southard, reviewing Kraepelin, would affirm that in dementia praecox masturbation was frequent, obstinate, and purposeless. In this conflation of concept and behaviour, he further asserted that Bleuler counted as schizophrenic cases of insanity of masturbation (Southard, 1914). Elsewhere, Cole's *Mental Diseases* (1913) would argue that masturbation was a symptom of dementia praecox. While sources such as Younger's *Insanity in Everyday Practice* would note of 'dementia praecox catatonica' that 'Masturbation may be a cause as well as a symptom' (1914, p. 96). Yet others continued to document and probe extensively the masturbation history of their patients (Kirby, 1912/1915). And others investigated disturbances of the sexual glands as an essential pathological condition in dementia praecox. This was helped by case notes in which masturbation had been recorded as symptomatically noteworthy in those so diagnosed (Mott and Such, 1922). [The latter authors note the attempt to cure one patient with a high falsetto voice of masturbation by blistering the penis. The patient had been diagnosed with dementia praecox (among other things).] Such was its pathological magnetism, masturbation was even imagined where it clearly wasn't. Jelliffe and White (1919) would see the constant spitting of mucus in schizophrenia as symbolic of semen.

Yet in parallel with such viewpoints, more moderate views had also been gestating. There was a growing realisation of the ubiquity of masturbation. And in 1901 notables such as Havlock Ellis had dismissed masturbation as no longer claiming serious attention. Furthermore, the social fear of masturbation was itself becoming subtlety repathologised. It was becoming internalised into the patient in what Greenacre could call a 'masturbation–fear-of-insanity complex' (1918, p. 200). In such circumstances, the fear of masturbation causing insanity, an idea once advocated by alienists, and possibly faithfully garnered from their teaching, was now preferentially seen as a 'complex' belonging to the unenlightened patient. And this perhaps explains why in 1929 psychoanalysis and treatment of masturbation in schizoids or schizomaniacs was seen as valuable—for those children caught in time (Robin, 1929). This internalisation, among other things, would help herald a full demedicalisation of masturbation per se in later years.

By 1962, the negative characterisation of masturbation had seemingly disappeared from all attempts to conceptualise madness. And E.H. Hare (1962) could safely announce the death of the idea of masturbatory insanity. The idea, rooted in a 1760 treatise by Swiss physician Tissot, was history. In doing so, Hare repeated earlier comments by Kraepelin that noted that the resident physician of the Royal Edinburgh Asylum, David Skae (1814–73), had explicitly declared a specific type of insanity due to masturbation. Kraepelin had argued that only for Skae's unlucky choice of nomenclature, Skae might be remembered for one of the earliest descriptions of Hecker's hebephrenia. In 1977, Ellard, citing Hare (1962), could consequently claim that 'The concept of schizophrenia began as the concept of masturbatory psychosis' (1977, p. 13). In other words, the embrace of schizophrenia and masturbation had now been formally reconfigured. It had been historically sanctified as a marker of a transhistorical concept/disease (Hare, Kraepelin). And perhaps even as an implied indicator of enlightened progress (Ellard).

Sexuality

Given that masturbation was pathologisable, it comes as little surprise that so too was sexuality and its expression. For Kempf, for example, 'The catatonic adjustment in males is due, except upon rare occasions, to the fact that the dissociated sexual cravings are perverse and require the reception of homosexual attentions' (1920, p. 557). Elsewhere we find the description of a case where 'a latent homosexual trend' came to the surface in a woman diagnosed with schizophrenia, 'which developed into the fantastic belief that she was a man (Anon, 1923, p. 462). Such arguments built naturally on earlier speculation by Bleuler himself who noted that, although unproven, it was not improbable that a relationship existed between catatonic symptoms and sexuality (Bleuler, 1911/1952). Bleuler would also note that 'other patients are in love with a ward-mate with complete disregard of sex ...' (1911/1952, p. 52). Similarly, Kraepelin, in a passage on sexual behaviour, had noted that 'Female patients are more apt to associate with their own sex' (Diefendorf, 1918, p. 235).

By 1940, Hinsie and Shatzky continued to draw attention to 'morbid concepts' such as homosexuality, which 'form the framework for the new and phantastic [sic] universe to which the patient adjusts himself' (1940, p. 150), while Patterson (1940) would cite from the literature that deep-seated homosexual conflicts resulted in a splitting of the personality with the development of paranoid dementia praecox. And, by 1974, following a raucous year that saw the APA concede that the pathologisation of homosexuality was problematic, other schizophrenia researchers would continue to speak of a failure to transition from adolescent to a mature heterosexuality (Calanca, 1974). While another schizophrenia study could deem as abnormal behaviour 'avoiding the other sex completely, homosexuality, promiscuity, or bizarre sexual indulgences' (Davis et al., 1974, p. 63).

Where it existed, prejudice could be sweeping and nonchalant. In 1974, John Money could declare that 'there are some who would see all transsexualism as schizophrenia or possibly as paranoia transsexualis' (1974, p. 347). In 1975, a depressed man's concern that other people were talking about him and his homosexuality was recorded as being incorrectly assessed as a paranoid schizophrenic delusion (Lehmann, 1975). While, in 1977, Joseph Berke noted that one young man had been diagnosed as schizophrenic—by his family doctor—because he had purchased Italian stiletto shoes and wished to become sexually active, over his parent's objections (Berke, 1977).

For a seemingly puzzled Ellard (1977), many schizophrenics presented as insecure in their sexual identity. Although, in fact, the real insecurity often lay in others. Schulz, observing that patients were uncertain over sexual identity, noted by way of example that a woman might use some phrases such as 'just like any man would' (1975, p. 56). As we shall next see accompanying such insecurity, one might also have found the fear of a lurking criminal propensity.

Sterilisation

Belief in the behavioural profession's ability to provide societal relief from its supposed problems was understandable. Psychiatry certainly seemed to possess sufficient ideological drive, classificatory power, and the necessary technology. Ostensibly, all this could partially calm the social worries it had often helped create (see Noll, 2015), or resolve the problems it had prophesised. A study by Joel T. Braslow of sterilisation practice in the name of therapeutics documents the following case occurring around the 1920s in Stockton State Hospital California:

> Dr. McCoskey: She cannot come. She has psychosis with somatic disease.
> Dr. Conzelmann: Why not call her something she can be sterilized by?
> Dr. McCoskey: She won't need it, she is going to die, she has been paranoid three years.
> Dr. Schreiber: Dementia Praecox with Tuberculosis.
> Dr. McCoskey: If she lives we can change the diagnosis.
> Dr. Conzelmann: Dementia Praecox Paranoid form with Tuberculosis, so we can sterilize her? (All agree) (1996, p. 37)

Sterilisation followed earlier calls for legal prohibition of marriage for premorbid personalities in dementia praecox by Smith Ely Jelliffe (Noll, 2011). It also followed debates concerning the financial burden of the insane (Eraso, 2010). Later, in 1923, Swiss lawmakers such as one Dr Hauswirth were not only openly advocating for castration or sterilisation of people diagnosed with schizophrenia, but even the death of incurable patients: 'Une autra solution serait la mise à mort des aliénés incurables et des idiots' (Nisot, 1929, p. 510). Such practices reflect an aspiration for patient sterilisation, which came into being with the concept itself. For even from the concept's beginning, castration was used in Burghölzli. And, as such, Bleuler himself speaking of schizophrenia had expressed a desire that 'sterilization will soon be employed on a larger scale ...' (1911/1950, p. 473). By 1930, *The Lancet* could report that Büchler of Budapest had argued that 'it is beyond doubt that the reproduction of schizophrenics is undesirable ... efforts must be directed to the prevention of gestation' (Anon, 1930, p. 420).

Doubtless, the stigmatising dementia praecox (and elsewhere schizophrenia) was able to function in conjunction with sterilisation as a perfect conduit for cleansing the future. It had diagnostic elasticity. But there are caveats. In 1930s Argentina, negative eugenic measures were *not* actively implemented. Instead, they gave way to coercive institutionalised labour (Eraso, 2010). Hence, it must be acknowledged the degree to which sterilisation (at times voluntary) could be supported by researchers was often nuanced. Françoise Minkowska, for example, is on record as having been 'hostile' to forced sterilisation around 1938 (Odier, 1939). That said, Minkowska did favour sterilisation under certain circumstances. Similarly, in Dowbiggin's investigation of North American psychiatry and eugenics there was virtually no psychiatrist who did not on occasion express an opinion favourable to eugenics. However, individual reactions from schizophrenia researchers to sterilisation were complex. Some, such as Adolf Meyer, for instance, would eventually conclude that eugenics was little more than an ill-disguised attempt by one group to subordinate another. William A. White feared that eugenicists were forever overstating their case, and might embarrass psychiatry. Ernst Southard, too, was impatient with eugenic pessimism. And, indeed, Meyer and White were only intermittently focused on strictly eugenic matters (Dowbiggin, 1997). At best, however, such doubts were not sufficiently amplified. And collectively, psychiatry, as a profession, helped to create a moral and cultural climate that nurtured eugenics. Furthermore, as Dowbiggin (1997) also notes, lawmakers listened closely to their recommendations.

Notoriously in 1933, Hitler's Third Reich introduced involuntary sterilisation, under 'The Law for the Prevention of Genetically Diseased Offspring'. It was then reified in so-called genetic courts that would affect close to 400,000 Germans, including those diagnosed with schizophrenia. There was nothing accidental about this. The law was drafted with significant contributions from psychiatrists with experience relating to dementia praecox and schizophrenia, such as Ernst Kretschmer and Ernst Rüdin. And, in particular, Rüdin's erroneous belief that a recessive Mendelian gene contributed to dementia praecox would be a basis for the eugenic and racist policies of Hitler's regime. Twice decorated by Hitler, he was labelled by a contemporary as the 'Reichsführer for Psychiatry and Sterilization' (Seidelman, 1988, p. 222).

The desire for sterilisation of schizophrenia found further expression in the work of Franz Kallman. For Kallman, schizophrenia was an unceasing source of maladjusted cranks, asocial eccentrics, and the lowest type of criminal offenders. Supposedly, they were neither able nor willing to make use of their individual liberty. As such, he sought 'The prevention of several hundred schizophrenic patients and their tainted descendants' (Kallman, 1938, xiii). Kallman, in his efforts to further eugenics, would speak of the disproportionate presence of 'such symptoms of schizoid abnormality as bigotry, pietism, avarice, superstition, obstinacy or crankiness ...' (1938, p. 103). He further argued that care for the weak and the diseased was *outweighed* by the obligation to protect the continuance of biologically sound families.

Depending on circumstances, Kallman urged restraint in institutions, prohibitions on marriage, and voluntary or compulsory sterilisation. Kallman called for the systematic genetic education for physicians, teachers, and social workers. A state archive of tainted families should be established. He also targeted siblings and relatives for intervention in various ways. In 1937, the concept of schizophrenia had been recognised by future French Resistance member Paul Schiff as too vast to lend itself to *serious* genetic studies (Schiff, 1937). Yet Kallman still made such studies possible by a series of nosological fudges. In his research, the necessary taxonomic synonymy was imposed on diverse and historic patient histories that predated schizophrenia's inception (some patients were born as early as 1820) (McNally, 2009).

Ultimately, in conjunction with the support and contribution of numerous other psychiatrists, such beliefs anticipated and constituted part of the intellectual apparatus underpinning the holocaust. Carl Schneider, for example, 'served as one of the most important experts in the sterilization and murder of the mentally ill until the intercession of the Catholic Church and Cardinal von Galen in 1939' (Gilman, 1985, p. 594). Schneider's research plans also indicated that histology tests would complete his research 'after the patient's (provoked) death' (Eraso, 2010, p. 73). By August 1941, a further 41,000 mental patients had been gassed in an even more radical 'euthanasia' programme. It is important to note that the original act did not order doctors to murder these patients. It only empowered them to do so (Cocks, 1997, 1994).

The horrors of the Second World War did not put an end to sterilisation. After the war, the issue of sterilisation in schizophrenia continued to be discussed in some quarters. Debate focused on the kind of technical challenges that a recessive gene might present viz. the number of generations that needed to be sterilised (Tétry, 1948). Furthermore, sterilisation of those diagnosed with schizophrenia and those thought to have 'inferior genes' continued to be introduced in some countries, for example, postwar Japan (then under American oversight). And it lasted in others, such as Sweden, right into the 1970s (Dowbiggin, 1997).

Racial and Ethnic Bias

In 1974, a cross-national schizophrenia study took place between locations in London and New York. It was discovered that American psychiatrists generally applied the diagnosis of schizophrenia to a much wider variety of clinical conditions than did their British or British-trained colleagues. Furthermore, on the American side, among other things, issues of racial bias also arose. The study spuriously attempted to explain this away as a 'social distance' effect; that is, the tendency of black patients to be suspicious, uncommunicative, and out of work (Anon, 1974). No charge of racism was made. Indeed, by this time the ancillary literature was growing increasingly focused on complex discussions concerning the causes of ethnocentric bias in diagnosis. There would be no shortage of data. In the 1980s in the UK, for instance, ethnocentric bias remained much in evidence. Irish, Afro-Caribbean, and Asian were all over-represented in groups diagnosed with schizophrenia (Coppock and Hopton, 2000).

Historically speaking, however, such findings have a long tradition and such biases appear to have been long reflected in certain institutional data sets. In 1925, Horatio M. Pollock concluded that 'Schizophrenia is more prevalent among negroes than among whites'. Pollock (1925/1928) had observed that in New York State the diagnosis rate per 100,000 was 48.6 among black people. Yet it was only 16.9 among white people. Leopold Bellak similarly reported a New York study by one Malzberg for the three years ending in 1931. In comparing the standardised rates among black people with the standardised rates among white people he gives the following figures: 'for Negroes [sic], 51.1 per 100,000; for whites 25.7—a ratio of 2 to 1' (Bellak, 1948, p. 15). By 1972, while schizophrenia was diagnosed in 51 per 100,000 white people admitted to state and county mental hospitals in the USA, the figure stood at 118.6 per 100,000 for nonwhite people (197.1 per 100,000 black males vs. 56.3 per 100,000 white males). 'Misdiagnosis' was readily recognised in the literature, 'due in part to racial biases' (Ruiz, 1982, p. 323) and not just confounding social factors. White people were more likely to receive a diagnosis of manic depressive. Why? Because 'it was believed that Blacks were too happy go lucky to be manic depressive' (ibid).

Elsewhere, in apartheid South Africa, 56% of the 400 male Bantu patients admitted to Weskoppies Mental Hospital Pretoria (between December 1952 and February 1954) were diagnosed with schizophrenia (Moffson, 1955). Such data were complemented in the literature by a matrix of scientific findings, ostensibly revealing racial differences. Strecker and Ebaugh, for instance, conducted a study on dementia praecox post-childbirth that noted that 'It seems fair to conclude that the mental stability of Jewish women is more prone to be upset by the stress of childbirth than that of other women' (1925/1926, p. 250). Similarly, in 1925, Brill could comment on the schizoid disposition:

> Negroes are decidedly syntonic, while the American Indians are preponderantly schizoid. Even among civilized races one can discern

> a preponderance of one or the other factors; the northern blond types, Nordics, are more schizoid than the southern darker races (1925/1928, p. 37).

Most likely, such 'scientific knowledge' was little more than prejudice opined. Later, even official field journals carried comments that simple racism had permeated whole chunks of the literature. In 1973, for example, Torrey would admit that the works of J.C. Carothers, 'which have been widely quoted in the epidemiological literature on schizophrenia, really are more appropriate as classical works of racism' (1973, p. 56). [Earlier, in 1940, Carothers had declared that 'the normal African is not schizophrenic, but the step from the primitive attitude to schizophrenia is but a short and easy one' (McCulloch, 1995).]

Newspapers, in turn, could report various such scientific findings. Hans Steck's *Psychiatrie et Biologie*, for example, was cited by the *Gazette de Lausanne*. The newspaper reported the opinion that schizophrenia was a regression to a mystical and prelogical primitive mentality—as described by sociologists in the tribes of 'savages' (Anon, 1927). Such 'knowledge' was seemingly authoritative. Kraepelin, for instance, had singled out Jews for their frequency of psychopathic disposition (Wyden, 1998). Other psychiatric authorities such as Richard Kraft Ebbing and Theodor Kirchoff had also done so (Gilman, 1985). And, again, such 'knowledge' would find ready utility in the practice of genocide. As when professor of ethnology George Montandon penned his *Le Matin* article '*Comment reconnaître les Juifs*' ('How to Recognise the Jews'). The article would charge Jews with higher levels of schizophrenia (Montandon, 1941).

Mother and Family

It is now well recognised that the family played a complex role in the negotiation of twentieth-century madness. But in schizophrenia research, family, like sexuality, could also be pathologised. Hence, in 1925, we can find social hygienists noting the importance of family conflict in the genesis of states described in schizophrenia (Toulouse and Mourge, 1925). This often occurred in a space where theoretical assumptions were easily passed off as fact. And it occurred in a space where parents, siblings, and offspring could be targeted for sterilisation (Anon, 1933). Even the simplest maternal 'failings' were deemed capable of, at the very least, inducing pathology resembling schizophrenia. In 1936, French psychoanalysts were informed in a conference that the frustration or disappointment of a child could contribute to the development of a schizophrenia attitude.[1] This occurred via regression to a state of autism, which could be mistaken for real schizophrenia (Leuba, 1936). At the same time, the term schizonoïa was being used to reference individuals whose development had been disrupted in early emotional relations with their mother. Typically, the disruption was caused by the mother (Laforgue, 1927). Most notably, such ideas culminated in the notorious term 'schizophrenogenic' being applied to the mother by psychiatrist Frieda Fromm-Reichmann:

> The schizophrenic is painfully distrustful and resentful of other people, due to the severe early warp and rejection he encountered in important people of his infancy and childhood, as a rule mainly in a schizophrenogenic mother (1948/1959, p. 164).

This term was also shortened to the 'schizogenic' mother (Nuffield, 1954). Proposed solutions could be traumatic. Falstein and Sutton, for example, proposed that 'in almost every instance, the schizophrenic child must be removed from its "sick" mother in order that adequate remedial measures may be instituted' (Falstein and Sutton, 1958, p. 667). Fromm-Reichmann's term was objected to as reprehensible as early as 1956 (Racamier, 1957).

By 1960 Laing could note that 'fortunately' an early witch hunt quality about the concept had begun to fade. Perhaps for this reason he still used it (Laing, 1960/1990). By 1972, however, the idea of the schizophenogenic or refrigerated mechanical mother, who produced psychotic children, was noted as not well supported (Hingtgen and Bryson, 1972). Yet, when not being considered overly seductive (qua *la mère séductrice*), mothers could still find their behaviour ineluctably pathologised on a spectrum of passivity to over controlling (Heuyer, 1974).

As noted, such thinking did not stop at the mother, but embraced the family itself. As such, Irving Kaufman and colleagues further classified parents, according to their personality structures. They could be 'pseudoneurotic', 'pseudopsychosomatic', 'pseudodelinquent', and 'overtly psychotic' (Kaufman et al., 1960). [Some types of 'schizophrenic personalities' around this period included the undifferentiated, inadequate, subparanoid, and schizoid (Gottesman and Shields, 1976).] Similarly, Laing (1960/1990) would speak of the schizophrenogenic family. While in 1973, B.B. Wolman could speak of the schizogenic family. This in itself was seemingly a throwback to a much earlier schizothymic family (Wolman, 1973). [Wolman (1966) had earlier described 'manifest schizophrenia' and 'vectoriasis praecox'.]

The pathologisation of family was, consequently, wide ranging. Indeed, W.B. James is credited with the aphorism: 'For every schizophrenic patient there is one schizophrenic parent' (Wilson, 1951, p. 1502). This could even be intergenerational, as when Lewis Hill asserted that it took three generations to make a schizophrenic (Ey et al., 1977). However, the supposed pathological traits in families were found not to be unique to families with a member diagnosed with schizophrenia. Nor did cases of schizophrenia always result from such supposed schizogenic families (ibid). As the century closed, Peter Wyman's *Conquering Schizophrenia* would symbolically register the theoretical change since the 1960s. It excused the field by stating that science had progressed (once again). And it exonerated parents of guilt in the process of triggering or encouraging mental illness in their kids (Wyden, 1998).

Political Dissent

Beyond ensnaring all of the above, schizophrenia classification extended into the overtly political. Bleuler, for example, could claim that 'a considerable part of army deserters are schizophrenic wanderers' (Bleuler, 1916/1924, p. 413), despite the recognised horrors of trench warfare in the First World War. Similarly, his assistant Jörger (1918) evaluated peace protesters in his asylum as schizophrenic or potentially schizophrenic (despite some showing an absence of symptoms). Elsewhere, French reports had also attempted to link simulation of mental illness in the military as indicative of schizophrenia (Livet, 1914). And later, supporters of De Gaulle and the French resistance were declared to be suffering a new illness known as dinguallisme. This was characterised by, among other things, schizophrenia (Allard, 1941). [n.b. Military life and combat would supposedly also result in schizophrenic war psychoses qua *schizophrene Kriegpsychosen* (Schneider, 1918), and 'the dementia praecox type of reaction in soldiers' (Anon, 1923, p.461). It also supposedly resulted in three-day (Eastman, 1945), four-day, and five-day schizophrenia (Brill et al., 1969; Kormos, 1978).[2]] Broad brushstrokes could easily be applied to whole nations that had politically erred. In 1959, for example, the Archbishop of Canterbury could single out the Japanese as suffering an acute form of schizophrenia. This was on account of their supposed indifference to human suffering during the war, possibly following similar comments made in Reverend Willis Lamott's 1944 earlier book *Nippon* (Anon, 1959).

Casual brushstrokes could also be applied to classes of politically deviant individuals. Most notably, in 1971 the world was alerted to the fact that hundreds if not thousands of Soviet political dissenters had been detained in Soviet asylums, which some would label psychiatric gulags (Porter and Micale, 1994). They had been diagnosed as suffering slow 'sluggish schizophrenia' (not exclusively—the category '*vialo-tekushchaia shizofreniia*' first formulated by Grunia Sukhareva (Zajicek, 2014) and linked with Snezhnevsky, had other uses). Symptoms included 'paranoid reformist delusional ideas' (Bloch and Reddaway, 1977; Fireside, 1979). Such threatening ideas seemingly included complex economic and social theories put forward as alternatives to orthodox Marxism (Wing, 1974). They also apparently included the urge to emigrate (Szasz, 2009). In essence, anything threatening appears to have been citable. Symptoms of composer Pyotr Starchik's 'creeping schizophrenia' included his religious beliefs and rudeness (Anon, 1981). Terms such as 'philosophical intoxication' were in use to describe the onset and early stages of the disorder (Ougrin et al., 2006).

Arguably, the idea of 'philosophical intoxication' did have a 'Western' counterpart. That is, in schizophrenia's so-called 'pseudophilosophers'. In such cases the presentation of schizophrenia could manifest as 'psychotic pseudoprofundity'. Ellard (1977) noted this was difficult to discern from esoteric beliefs—and *sometimes true*. Such conceptual scope then existed as a lurking threat to those who held so-called esoteric beliefs. It existed in a time when many in Western society feared the use of psychiatric diagnosis on dissidents outside of Russia (Gelinas, 1977). And it did so at a time when members of the Campaign for Nuclear Disarmament were alleged to have been sent to mental hospitals instead of jail (Berke, 1977). The level of such suspicions and fears should not be underestimated. In 1956, for example, none other than a Russian ambassador had claimed that healthy Russian citizens were being detained in German asylums and that they had been made subject to medical experiments using shock treatment. This prompted an official response from Germany to deny that Russians were guinea pigs (Anon, 1956).

Against such events, conceptual doubts continued as usual. In 1972, consultant psychiatrist, registrar, and clinical psychologist, David Shaw, Sidney Block, and Ann Vickers, respectively, reported on Soviet 'Psychiatry and the State'. They argued in the *New Scientist* that there was no firm evidence to warrant the term schizophrenia in Russia (Shaw et al., 1972). As part of their analysis, the trio attempted to skewer Russian psychiatrist A.N. Snezhnevsky. They based their attack on his prior admission that schizophrenia lacked a unified classification. For these critics, the Soviet classification appeared totally arbitrary (one type of schizophrenia—the 'complex galloping syndrome'—was attributed to Snezhnevsky himself). There existed frequent confusion between syndrome and forms of the disease. And there also existed confusion between syndrome and symptoms (ibid). Yet the Russian situation was clearly a cause for reflection on 'Western' psychiatry. Describing themselves as part of a 'significant minority', Shaw et al. noted 'confusion and imprecision in the diagnosis of schizophrenia in particular' (1972, p. 259). They cited, approvingly, L.S. Kubie, who had declared in 1969 that at last a rebellion against the concept of schizophrenia was in full swing. [For a more comprehensive account of Russian and Soviet schizophrenia conceptualisation, which differs from the West, including Ukranian 'slow-flow' schizophrenia, a variant of paranoid schizophrenia, see Lavretsky (1998).]

The capacity for schizophrenia to function as a catch-all for political dissent appears to have spanned much of the century both in theory and practice. As late as 1981, for example, political scientist James Clarke critiqued attempts by psychiatrists to explain historical attempts to assassinate North American presidents. The explanation of choice was schizophrenia, and, on occasion, simply paranoid schizophrenia. For Clarke (1981) such 'repeated' and 'erroneous efforts' could only be achieved through gross and inaccurate generalisations. In the case of Soviet dissidents, events reached an ostensible closure in 1983, when Soviet psychiatrists, at risk of suspension or expulsion, quit the World Psychiatric Association (Herbert, 1983). However, it is not clear that such practices ever ended in the twentieth century. By 1991 dissidents continued to be detained in Ukrainian psychiatric institutions (Ougrin et al., 2006). Similar occurrences have taken place more recently in other places, for example China, the psychiatry of which was partially influenced by Russian thinking (Appelbaum, 2001). However, in these later cases, the precise 'diagnostic' use of schizophrenia remains to be fully assessed.

Conclusions

It should be stressed that in day-to-day psychiatric practice schizophrenia does not appear to have been used as a means of social control. Schizophrenia was a versatile and extensible concept. But, by and large, it was not consistently and indiscriminately used to frame gender or some particular social deviance regarding, say, political behaviour. By contrast, the identification of homosexuality in the APA's DSM-II as a 'sexual orientation disturbance' probably was used as a tool to regulate behaviour. When applied to diagnosis in practice, the collected phenomenology and conceptualisation of schizophrenia did not function is such a way.

Moreover, any attempt to put a percentage on the absolute number of cases of schizophrenia that were exclusively the result of cultural prejudice seems like an impossible exercise. Such prejudice, although present in a global sense, becomes murkier at the level of individual analysis. Things are often unclear in case books, where a plausibly victimised individual is ensnared by lists of additional classical symptoms. There is frequently no way of knowing if such classical symptoms were, or were not, the product of the psychiatric imagination. Prejudice could also operate in conjunction with mental illness. And it is inconceivable that mental disorder qua one or more diseases, did not affect nearly all categories of people in some way.

Nevertheless, the articulation of prejudice within and around the schizophrenia literature is easily found and, in fact, did not go unnoticed. In 1925, French surrealists, referencing dementia praecox, attacked psychiatry for a hundred pretentious pathogeneses. They criticised it 'for a hundred classifications of which only the vaguest one are all usable' (Roudinesco, 1986, p. 7). By 1930 their leader and former psychiatrist André Breton would condemn Bleuler's autism as an abusive attack on forms of desertion, refusal, and disobedience. Similarly, by 1959, John Bartlow Martin's *The Pane of Glass* would peel some of the veneer off schizophrenia:

> Kovitz smiled. 'The terms don't fit. Our classification of schizophrenia is absolutely unscientific. But it doesn't matter. He is purely a custodial problem—there is nothing to do but take care of him' (1959, p. 28).

For Martin's Kovitz, schizophrenia classification was deemed 'absolutely unscientific'. But in all likelihood such statements emanated from psychiatrists themselves. This particular attempt at literary realism was grounded in the results of an investigative account of the workings of Columbus State Hospital. The work also noted psychiatry's ideological drive to accommodate society's wants.

Without clear boundaries and biological markers the ensnaring of the morally insane certainly appears to have been an easier task than might have been otherwise. Schizophrenia was a flexible concept. Indeed, we have examined here only the more notable trends. And it should be pointed out that multiple rarer examples of social prejudices can be found in the historical literature. These are ones that present-day researchers let slide with a knowing wry smile. For instance, Kraepelin's symptomatology noted patients speaking *shamelessly* about sexual matters (Diefendorf, 1918). Similarly, in one instance in 1930s Russia, a case of shyness was seemingly enough to acquire a label of schizophrenia (Zajicek, 2014). Perhaps, however, because they represented competing sources of knowledge, books and reading garnered particular attention. In 1925, for example, Kretschmer noted that 'Schizoid men, even of lowly origin, are generally lovers of books and nature, but it is with a certain eclectic accentuation' (1925/1999, p. 164). Hence, Evans, citing Kretschmer, noted that '*Certains schizoids sont ... amis des livres*' (1950, p. 250), while a drawing by Curran and Partridge in 1955 later shows the asthenic schizoid holding his book. In 1960, an ethno-psychiatric investigation in Ghana by Margaret Field also linked schizophrenia with the mere *intention* to use a grimoire. Hence the claim that 'the intention to make magic with *The Sixth and Seventh Books of Moses* frequently herald the onset of schizophrenia' (Field, cited in Elkins, 1986, p. 216). In 1975, the French concept of 'discordance schizophrénique' included the key symptom *bizarrerie*—paradoxicality that disturbed the observer; examples of *bizarrerie* included detachment and impenetrability, and the obsessive delusion of wanting to possess a typewriter (Ey et al., 1977). And Andreason could declare that although James Joyce never became schizophrenic his art *did* (Hare, 1987).[3]

Such statements never constituted a significant body of opinion within the schizophrenia literature. But they exist. And they often dealt with moral uncertainties which psychiatry, like society, was unfamiliar with. In 1921, for example, with Western society's growing engagement with 'Eastern' culture, Walter Lurje would declare that many details of the life and teaching of Buddha represented what those diagnosed with schizophrenia do and think (Lurje, 1921). Similarly, in 1972, with growing public interest in the occult, psychiatrist Lawrence Kayton attempted to associate the 'vampire legend' with schizophrenia. Kayton wrote, 'With the growing interest in vampires and other such monsters, an increased incidence or visibility of schizoid and schizophrenic problems may be manifesting itself' (1972, p. 304).

If we were to give them unity, we might say that these localised expressions often manifested suspicion towards the outsider or newly emerging acts of cultural expression. In 1944, Curran and Guttmann would claim that 'The dreamy, bearded, be-sandalled [sic] denizen of Bloomsbury or Montmartre is often a schizoid' (1944, p. 59). In 1974, psychiatrists could further ask how many abstract painters and surrealists resembled schizophrenics in their emotional tendencies and uncontrolled dreams (Heuyer, 1974). Similarly, for Schneider, hebephrenia had often carried with it 'pathoplastically' the features of the period, such as the 'bobby soxer' (e.g. teenage girl, ardent Frank Sinatra fan) (Schneider, 1959). Clearly, in such cases, the concept offered solutions for society's anxieties and increased the cultural power of those who made such pronouncements.

Such framing may have been more important that one might suspect. We must remember that as late as 1978 Stephens could write that diagnosis depended almost entirely on the individual psychiatrist's definition of schizophrenia (Stephens, 1978). And so by that we may include in such definitions the implicit and unstated moral dimensions that guide each psychiatrist's assessment of a patient. Quite conceivably, the collation of such examples—even as singularities—would demonstrate an even more disquieting capacity and flexibility in the concept's ability to assimilate society's prejudices and fears, over and beyond the more striking themes charted above.

Twentieth-century schizophrenia classification then permeated and was permeated by the attitudes, values, and beliefs of society. It supported the affirmation of the morally insane. And, indeed, it did so while simultaneously facilitating the specific conceptual, bureaucratic, and ideological needs of the psychiatrists and other professionals who wielded it. In many senses, we appear to witness a profession whose insecure theorisation was presented for social affirmation and a concept in search of social utility. Nevertheless, for most of that time, schizophrenia research and conceptualisation trundled on, at best oblivious and at worst nonchalantly. It did so without every really knowing that which was increasingly at stake. This was, to paraphrase Roudinesco (1986), the status of a concept capable of facilitating the existence of such a perversion. And for some psychiatrists in particular, the more history realised that schizophrenia catered for society's prejudices, the less and less convincing were claims that it existed, in a strong way, as a scientific object independent of social context that shaped and maintained it.

8
Contesting Schizophrenia?

We have seen that problems with the concept of schizophrenia did not go unnoticed. We also saw earlier how surrealist André Breton had condemned Bleuler's autism as an abusive attack on forms of desertion, refusal, and disobedience. [Similarly, by 1924 Dadaist Hugo Ball's *Sieben schizophrene Sonette* further uses the schizophrenic as an exotic device to critique society (Gilman, 1985).] Other reactions could be more pointed. In 1924, for example, Bleuler's concept was slammed as bizarre, simplistic, vague, arbitrary, and insufficiently comprehensive (De Fleury, 1924). But, for the most part, such early tensions ruffled few feathers within psychiatry. In official psychiatry, acknowledged problems with the conceptualisation of illness remained, at best, an abstract intellectual concern. In 1923, for example, Kraepelin's successor, Oswald Bumke, cast doubt on the reality of dementia praecox asking, 'What if dementia praecox simply did not exist?' Yet Bumke's resolution to his own question was simply to express a preference for the term schizophrenic reactions (Noll, 2011).

Still, over the course of the century, dissenting voices slowly accumulated and amplified within the literature. This criticism reached a crescendo in the period traversing 1960's counterculture (or c. 1955–75). By now the debate often stirred public consciousness. And thanks to self-confessed antipsychiatrist David Cooper, the more unorthodox critics, such as R.D. Laing, would acquire the incorrigible label antipsychiatrist and their activities would be characterised as antipsychiatry (an appellation much denied). Antipsychiatry had variously been called meaningless, open to interpretation, or a point of convergence (Postel and Allen, 1994). Yet, tellingly, a significant number of the concept's critics—both orthodox and unorthodox—continuously emerged, either directly or indirectly, from the behavioural professions themselves—that is, from psychiatry, psychology, sociology, and so forth. R.D. Laing was a psychiatrist, as were David Cooper and Thomas Szasz. Even Foucault trained as a psychologist (and André Breton was a former psychiatrist). The psychiatric critic was, in effect, the psychiatrist's essential other.

The opening up and intensification of this social and conceptual criticism, from the early twentieth century onwards, took a number of forms. In line with orthodox criticism, some arguments exposed the concept's contradictions, and drew attention to its expansiveness and its inconsistencies. Other arguments promoted the idea that schizophrenia was, in one sense or another, a social label or ideological construction and, as such, rejected it. Yet others romanticised madness. Many arguments struggled to gain any sort of official acceptance (although they often reached a wide and sometimes sympathetic audience). And self-evidently, criticism that dismissed schizophrenia was rejected by mainstream psychiatry.

Nevertheless, criticism was functionally important because it held up a mirror to psychiatry regarding its weaknesses in formulating schizophrenia. Furthermore, it exposed internal professional disagreement over schizophrenia's conceptualisation to a wider audience (both in the sense of other disciplines and the wider public). And it is perhaps no coincidence that in an increasingly hegemonic North America, this period of sustained and vocal criticism ultimately foreshadowed a conceptual overhaul of schizophrenia in the DSM-III of 1980 (see Chapter 9). For, at the very least, such exposure necessitated justifying the concept of schizophrenia. It had become impossible for psychiatrists and their colleagues to simply tolerate the conceptual ambiguities of their arch concept.

It remains notable that post-1950 attacks on the concept of schizophrenia were accompanied, in a small number of places, by a number of highly visible and seemingly radical changes in psychiatric praxis and patient–practitioner relationships. The most striking and effective example was Franco Basaglia's powerful attempt to destroy the mental hospital as a place of institutionalisation, as symbolically embodied in Italy's 1978 'Law 180', which sought to replace psychiatric hospitals with community services. Nevertheless, advocacy movements for patients had long existed, as had doubts over the merits of asylums. In 1926, for example, Hans Maier (then working in Burghölzli) could argue that interning people with schizophrenia in an asylum was, in principle, harmful. It separated them from their exterior life. And, where possible, he and his colleagues lodged patients with farmers or family (Bleuler and Claude, 1926/2001). Similarly, significant protest movements can be found in places such as nineteenth-century France and Germany. And, as noted earlier, Cooper's adoption of the term *antipsychiatrie* follows an earlier usage by Bernhard Beyer in 1908. Alternative therapeutic programmes were equally present, for example at St Elizabeth Washington from as early as 1914 (Noll, 2011) and at Sheppard and Enoch Pratt Hospital in the 1920s (Mosher, 1999). Post-1950, the Philadelphia Association's Kingsley Hall therapeutic community and other centres opened as alternative spaces for people undergoing psychotic breakdown. So what was happening alongside deeper conceptual probing, in at least one sense, was a rediscovery and reconfiguration of old approaches and traditions, with updated ideas.

Perhaps this was, to some extent, inevitable. Even for orthodox voices, Erving Goffman's explosive exposure of the malign effects of the total institution (in his 1961 *Asylums*) was described as numbing but beyond doubt in its generality (Wing, 1961). Goffman had also revealed that for some asylum staff schizophrenia was a vague and doubtful syndrome title used more for hospital census needs (Goffman, 1961). But, principally, Goffman's work was a clear articulation of a period of long unease with institutions. It gave expression to widespread but scattered concerns. Arguably, something had to give.

Yet, despite the apparent radical change, Szasz would later assert, with some justification, that 'antipsychiatrists' did not reject the idea of mental illness. Nor did they abandon coercion practised in the name of 'treating' mental illness (Szasz, 2008). And, indeed, in the case of the Kingsley Hall, at least, it could be argued that it further collapsed into an anarchic and sometimes abusive institution. Nevertheless, the situating of patients beyond the asylum walls and attempts to deconstruct madness further widened social debates concerning schizophrenia's conceptualisation. And it is worth acknowledging that resituating patients brought the concept of schizophrenia out of dusty archives and conferences and into public focus. This was something acutely magnified by the stellar media status of people such as Basaglia (Shorter, 2005) and Laing, as well as patients like Mary Barnes. But, as stated, what we will focus on here is the actual critical debate concerning conceptualisation itself. So let us step back and explore the gradual amplification of schizophrenia criticism, as it unfolded in the twentieth century. Such an exploration will make explicit the kinds of criticism that we have hitherto only noted in passing as we attempted to gain an understanding of classification, definition, and so forth. Gathering such criticism moves our analysis beyond criticism that might be dismissed as minor quibbles concerning, definitions, symptoms, wording, or subtypes. And its evaluation reveals some of the more fundamental ontological doubts that plagued many of those who sought to conceptualise twentieth-century schizophrenia.

Unease

The main focal point of early debate concerned boundaries and classification. By 1924, in North America, a modicum of unease with schizophrenia classification can be found in the writing of Harry Stack Sullivan. The influential psychiatrist noted sarcastically how research workers 'must collate and classify their data seeking always the *fundamentum divisionis*' (i.e. the arch principle according to which a genus is divided into species) (Sullivan, 1962, p. 7, original emphasis). In doing so, Sullivan further noted the striking 'frequency with which the workers have passed early from the science to the philosophy of schizophrenia' (1962, p. 8).

Driving Sullivan's ire was his observation that recovered patients tended to be diverted from 'the praecox group' into concepts such 'as the benign stupor of Hoch, and nonpraecox catatonia' (1962, p. 7). This reclassification irritated Sullivan. It effectively meant that any positive treatment outcome—facilitated by the kind of psychotherapy engaged in with dementia praecox by Sullivan—ran the risk of being declared an issue of misdiagnosis. If therapy failed, the classification was dementia praecox. If it succeeded, it was something else. [The same complaint can be found as late as 1992 in Whitaker's *Schizophrenic Disorders* (1992). Similarly, so-called 'spontaneous recovery' was also declared as neglecting the role of nursing (Biddle, 1949).]

Sullivan was acutely aware of practical and philosophical problems concerning classification in psychiatric discourse. Yet Sullivan made a distinction between schizophrenia and dementia praecox:

> Schizophrenia, in the light of clinical observations, is not to be regarded as a primary disease such as that which one may visualize when mentioning dementia praecox (1962, p. 11).

In addition, he further proceeded to discuss his investigations of patients diagnosed elsewhere as 'hebephrenic dementia praecox'. These he saw to be actually made up of hebephrenic and 'deteriorating' schizophrenics—which he saw as containing many 'paranoid praecox' cases. Finally, he wrote of 'catatonic dementia praecox' being the category to which all schizophrenic psychoses should be allocated. It was the group where therapeutic endeavour was consistently encouraging.

Sullivan also appears to have been content to use any system that helped him illustrate the ideas and processes that he felt to be at work in schizophrenia. Meyer, for example, had elsewhere spoken of schizophrenia as manifesting behavioural reactions that he grouped, in the 1920s, as 'parergasia'. This was a term that he hoped might replace schizophrenia (Meyer, 1957; Noll, 2011). Hence, by 1938 Sullivan could also speak of a 'classical psychotic state', which he referred to as 'catatonic parergasia [sic]' (1938/1965, p. 82). Sullivan saw catatonic parergasia as a lighter form of personal maladjustment to the more fully developed form found in patients diagnosed with schizophrenia or dementia praecox (1938/1965). Sullivan's lurking classification concerns were consequently accompanied by a contradictory and competing professional requirement to continue communicating results through classification. He seemed unable to remove himself from the prevailing psychiatric discourse. Instead, he remained entangled in the incessant redistribution of patients through categories relating to dementia praecox and schizophrenia.

A Nosological Fiction

Sullivan's unease with schizophrenia's classification was not idiosyncratic. Even self-confessed nosologists such as Smith Ely Jelliffe—who valued the potential pragmatic and therapeutic implications of nosology—could shower both schizophrenia and classification with healthy scepticism:

> we must also recognise, as nosologists—which before noted must be regarded as a useful fictional, logical tool (Vaihinger)—what position may we assume with reference to that infinitely less precise nosological fiction which we term schizophrenia? (1927, p. 417).

For Jelliffe, schizophrenia was an even *less precise nosological fiction* than the fictional, logical tool of nosology itself. Even so, Jelliffe refrained from arguing for radical change. The perceived utility of classification outweighed its possible product: fiction (which does, after all, have its uses). Others were less damming but equally unsure of their nosological faith. In 1932, W.B. Philsbury would admit 'The classifications of insanity have been various, and there is still lack of complete agreement. The classification is at best a compromise ...' (1932, p. 211).

Continued criticism could still be found a decade later. In 1934, for example, Price's *A Textbook of the Practice of Medicine* would critically survey dementia praecox and schizophrenia conceptualisation:

> By some high authorities the concept of Dementia Praecox has been much elaborated, and a large number of sub-divisions made; but the distinctions drawn are with some difficulty appreciated, even by those used to the soaring flights of psychiatric abstraction. On the other hand, Dementia Praecox, Paraphrenia and Paranoia tend to be included under the term schizophrenia ... (1934, p. 1825).

In a sobering assessment of the terminology under discussion Price further added:

> The distinction in practice between paranoia, paraphrenia and paranoid dementia praecox is often of extreme tenuity; nor do descriptions found in the literature present anything like clear pictures of decidedly separate morbid entities (ibid).

Yet despite drawing critical attention to the confusion, Price nonetheless also drew back and proceeded in the usual manner of trying to draw distinctions between groups. The 'paraphreniac' [sic], for example, 'in contrast to the patient suffering from paranoid dementia praecox, does not show the emotional vacuity which is so marked a feature in the latter' (Price, 1934, p. 1832). Around the same time Lewin also retreated from the quasi-rhetorical observation that the whole dementia praecox–schizophrenia concept was now a matter of history. That henceforth it was heuristically sterile (Lewin, 1934). So although by 1934 researchers were increasingly critically aware of the problems of applied classification, they appear locked within this groupthink. A volley of criticism was followed by renewed attempts at subdivision or classification.

By contrast, others, while admitting difficulties in diagnosis and acknowledging differences between countries and schools within the same country, thought the differences slight and perhaps resolvable through considering the duration of symptoms (James et al., 1937). And yet others, probably in allusion to Meyer's school, could mount attacks on the, 'foolish statement ... sometimes made that diagnoses in psychiatry are futile labels, not worth the affixing' (Larkin and Gillies, 1938, p. 385). Such comments were voiced with the knowledge that the boundaries of schizophrenia were set 'generously wide', but that it would 'be necessary for each country, and perhaps each clinic, to work the statistical prognosis for the cases which it has agreed to classify as "schizophrenia"' (ibid). Assessing prognosis in schizophrenia was not like casting a horoscope or reading the entrails of a sacrificial beast. Diagnosis was of the 'first importance' (ibid).

The problem of conceptual confusion, however, was not going away. In 1936, Nolan D.C. Lewis had claimed that there were schizophrenic types but that there was 'too little knowledge to enable us to make scientific differential diagnoses' (1936, p. 36). By 1939, Sullivan responded to the claim and in doing so appeared disillusioned with the whole project of schizophrenia classification. Sullivan cited the passage as an example of the 'sad state of psychiatric thinking' (1939/1953, p. 150). Sullivan asserted that there are 'no types of schizophrenia, but only some rather typical courses of events that are to be observed in schizophrenic states' (ibid). Now forging ahead with his interpersonal relations approach to understanding schizophrenia, Sullivan critically commented:

> Note the absence of consensually validated thought in the pronouncement. All that is evident is the author's conviction that mental disorders have fundamental types (1939/1953, p. 151).

Arguably, Sullivan's thoughts were not atypical of senior researchers who, throughout their careers, had watched schizophrenia classification repeatedly stumble. However, the remark also had a certain prophetic character. The future of schizophrenia nosology would depend precisely on consensual thought as manifested through APA committees (albeit with little success on what constituted validation). And it is with such context, and seeming suspicion towards decree, that R.G. Hoskins (1946) would ask if schizophrenia was merely a semantic convention and an entity by fiat.

Post-1950s Critics

The APA's DSM-I emerged in 1952, and would continue in series to the present day (more of this in Chapter 9). However, post-1950 orthodox psychiatry would continue to expose the concept's deficits (although it did not, of course, abandon the concept). In 1956, for example, Ivan F. Bennet, of the Psychiatry and Neurology Service and Veterans Administration Washington, could uncontroversially declare, 'in the symptom constellations that we call schizophrenia ... we have here a wastebasket diagnostic classification and I think we are more interested in the effects of the drug upon certain symptoms that may be present' (1956, p. 415). In 1959, leading biochemical theorist Seymour Kety would also remind readers that 'a pathological lesion characteristic of schizophrenia or any of its subgroups remains to be demonstrated' (1959, p. 1528). In the same year an international conference 'Field Studies in the Mental Disorders' would show schizophrenia conceptualisation to be in disarray (see Chapter 9) (Zubin, 1961), while, by 1960, a leading authority such as Rümke would note numerous contradictions in the concept of schizophrenia. To some, for example, it was an entity, for others a syndrome. For some it was characterised by primary symptoms; for others there were no primary symptoms. For some it was organic; for others psychogenetic. For some it transitioned into normality; for others not. For some genetic factors were of utmost importance; for others they were minimal (Gelinas, 1977).

It is has to be understood then that it is within a context of long-voiced criticism that Thomas Szasz would make his famous 1957 contribution. For Szasz, schizophrenia had become an explain-all, a *panchreston*. It filled a scientific void (Szasz, 1957). David Cooper would later declare that Szasz's approach viewed schizophrenia as a bad attack of what Wittgenstein called the bewitchment of our intelligence by language (Cooper, 1970). But these initial comments made by Szasz on schizophrenia were relatively uncontentious.

They were particularly uncontentious in comparison with his subsequent over-reaching ideas concerning the myth of mental illness (Szasz, 1960, 1976). Szasz was not yet declaring that schizophrenia was the sacred symbol of psychiatry. Nor was he producing the so-called 'antitheory' (Annitto, 1977), which would result in attempts to fire him. He was merely giving a more pointed expression to a general dissatisfaction with the concept.

Cultural Backlash

As we have seen, criticism of the concept of schizophrenia was nothing new. Nonetheless, the second half of the twentieth century witnessed a period of more heightened and striking criticism. It is worth taking a little time to understand this intensity. After the Second World War, a broad cultural backlash to psychiatric practice, particularly relating to institutionalisation and treatment, was now also beginning to unfold. This occurred within the context of a wider intergenerational questioning of received Western traditions, norms, authorities, and governance. Consequently, questioning and reconfiguring notions of sanity and insanity became symbolically important. Such considerations were part of a critical articulation of a post-holocaust world, a world whose faith in rational governance was profoundly shaken.

In this critical generation, many came to accept the notion of those diagnosed with schizophrenia as victims of culture itself. In the seminal 1955 poem *Howl*, Allen Ginsberg saw the best minds of his generation destroyed by madness. Ginsberg's own mother had been lobotomised after Ginsberg helped sign the papers in 1948, while Jack Kerouac had been diagnosed with dementia praecox (Miles, 2002). Others, believing that schizophrenia could be provoked by cultural pressures such as baccalaureate preparation, claimed people with schizophrenia were martyrs (Dora, 1963). Yet other readings, suspicious of the family, would re-evaluate Freud's neurotics as schizophrenics. In the process, the possibility of overt incestuous material in the background of schizophrenics would be suggested (Glueck, 1963). Many would increasingly associate this generation with sexual liberalisation and new music. Yet, in Switzerland, what characterised this 'exceptionally gifted new generation' had already been summarised by 1963. Pierre Furter (1963) proclaimed it as the desire to shake structures, pacifism, antimilitarism, and criticism of schizophrenia.

Many in this generation were supported by ample accounts of leucotomy, lobotomy, tranquillisation, violence, and neglect in twentieth-century asylums. This was sometimes personally witnessed or reinforced by the legacy of popular books such as Albert Deutsche's *Shame of the States*. At other times it was given breath by late-1940s movies such as 'Snake Pit' and 'Bedlam' (Decker, 2013). Such ideas were further supplemented by the belief (exaggerated but understandable) that all chronic mental deterioration was the product of institutional life (Klerman, 1977). Many professionals shared variants of this outlook. In July 1967 the 'Dialectics of Liberation Congress' would seek to demystify violence in all its human forms. Leon Redler, Joseph Berke, R.D. Laing, David Cooper, Gregory Bateson, and Ross Speck would all attend (Erving Goffman withdrew) (Dialectics of Liberation, 2012). Consequently, many optimistically embraced reformist ideas (Decker, 2013) and a broader rehumanisation of psychiatry (Murray, 2014).

One further phenomenon probably helped cement all this. Post-1950, scientific and cultural debate over drugs such as LSD and mescaline and schizophrenia had come to the fore. Interest in mescaline and schizophrenia extended, at least as far back as Klüver (1926). And with respect to the scientific consideration of LSD, a serotonin hypothesis existed as early as 1951 (Feldstein et al. 1958). For various reasons, both drugs would be countercultural favourites. The intense phenomenological experience of both LSD and mescaline, at times intellectualised by notables such as Aldous Huxley, meant that the debate would extend beyond mere biochemistry. Osmond and Smythies (1951), for example, would declare that language was not designed to describe the weird world of the mescaline-taker and the schizophrenic. The effects of mescaline varied 'as much but not more than the symptoms of schizophrenia, which ... allow a wide range of variability' (Osmond and Smythies, 1951, p. 603).

With such an attitude, the use of mescaline or 'the insanity producing drug' LSD (Anon, 1957b) suggested for many that they had gained privileged insight into some of the symptoms of schizophrenia.[1] Accompanying this belief, 'cannabis culture' was sometimes speculated as being responsible for declining schizophrenia hospitalisation, and noted to have been 'appreciative' of 'schizoid ideation' or 'lateral thinking' (Hasleton, 1974, p. 4). Similarly, Loren Mosher (1999) could explicitly state that interpersonal phenomenological interventions were akin to being an LSD trip guide. All this helped set the stage and background for greater conceptual revision in schizophrenia research than might otherwise have been the case. At least temporarily, madness seemed transparent. Although for some establishment psychiatrists, expressing and 'acting out' the nonconformist ideology of the 1960s counterculture was itself evidence for schizophrenic pathology (Lehmann, 1975).

More Sane than Mad

Post-1950, the boundaries of madness seemed increasingly unclear. Even in traditionally conservative countries, orthodox members of society can be found calling the sanity of society into question. Hence, by 1959, even the Archbishop of Canterbury, would opine that everyone was a little mad. Every nation was tainted with schizophrenia of one sort or another (Anon, 1959). Such thinking was not new. Even Rüdin and Sullivan had complained—the former seriously, the latter ironically—of the difficulty in finding the 'non-schizophrenic' (Bleuler and Claude, 1926/2001; Sullivan, 1962). Moreover anti-establishment thinking now saw this viewpoint as a way of obscuring the relationship between sanity and madness.

As Foucault would later argue, 'Nobody is more conservative that those people who tell you that the modern world is afflicted by ... schizophrenia. It is in fact a cunning way of excluding certain people or certain patterns of behaviour' (1974, p. 188). Hence, instead, what now found expression—and as exemplified in R.D. Laing's *Politics of Experience*—was the notion that the mad are sometimes *more sane* than the normal (Laing, 1967b). Such thinking was rooted in the observation of communication within families and expressed in Laing's earlier existential and phenomenological explorations in *The Divided Self* (1960/1990). This was a work that sought to make madness and the process of going mad comprehensible. In it, Laing noted 'the cracked mind of the schizophrenic may *let in* light which does not enter the intact mind of many sane people whose minds are closed' (1960/1990, p. 27). Laing felt the schizophrenic ceased to be schizophrenic when he met someone by whom he felt understood. Laing (1960/1990) attributed the idea to Jung but noted it was also sometimes voiced by Laing's patients.

Even more romantic variants of such ideas also emerged. David Cooper, for example, had once thought that 'schizophrenics were the strangled poets of our age' (1967, p. 109). (By contrast, Hilde Bruch would assert that the poet is a master of language, the schizophrenic a slave to it.) Elsewhere, Julian Silverman, following Ackerneckt in 1943 (Noll, 1983), argued that the essential nonparanoid schizophrenic form is regarded as more comparable with that of the shaman, the 'healed madman' (ibid; Silverman, 1967).

Such wishful thinking could sometimes obscure more serious arguments. Schizophrenia, for Foucault, in his 1961 *Histoire de la folie à l'âge classique—Folie et déraison* was an extension of prior taxonomies. These encompassed the discursive field he termed unreason. Of such taxonomy he noted 'it is as though these classifications had been an entirely empty activity, unfurling itself to find nothing at all, constantly being corrected in vain, a ceaseless activity that never succeeded ...' (Foucault, 1961/2006, p. 195). [An impressed Laing and Cooper helped introduce Richard Howard's 1965 English translation (Burston, 2001).] Yet even Foucault would speak of a madness with 'inaccessible primitive purity' (1961/2006, p. xxxiii),[2] although, as Hacking notes, he later supressed the comment (Foucault, 1961/2006). For all that, romanticism was not a core theme in this period.

More notable were the numerous philosophical influences converging on this radical reconceptualisation and questioning of madness, most notably strains of existential psychology and phenomenology. These were often rooted in the works of thinkers like Jaspers, Bergson, Husserl, Heidegger, Sartre, and Hegel. Hence, for Boss, schizophrenia was seen as an exemplar of the crisis of the sciences of psychopathology and phenomenology and offered a route to exit this crisis (Gelinas, 1977). The *British Medical Journal* (*BMJ*) speculated that such an orientation was a possible throwback to an earlier period when psychiatry was dominated by philosophy (Anon, 1957a). Yet, at least outside of Anglophone countries, this leaning was not all that unorthodox. Indeed, it reflected the mid-century fascination a large number of European psychiatrists (e.g. Minkowski, Binswanger) had with existential analysis. Hence an observation that antipsychiatry had replaced statistics and verifiable methodologies with ontological criteria was not really a criticism. It was a statement of fact (Gelinas, 1977).

Romanticism and philosophy aside, a core critical stance towards schizophrenia remained. As early as 1964, Laing and Esterson, in *Sanity, Madness and the Family* (1964/1970), had, citing Szasz, transitioned to the viewpoint that to regard schizophrenia as a fact was unequivocally false. For them, a study of 25 'schizophrenic' families from 1958 to 1963 (facilitated by Tavistock's John Bowlby) had facilitated such a declaration. They noted, 'In our view it is an assumption, a theory, a hypothesis, but not a *fact* that anyone suffers from a condition called "schizophrenia" ... We do not accept "schizophrenia" as being a biochemical, neurophysiological, psychological fact' (Laing and Esterson, 1964/1970, pp. 11–12).

Schizophrenia: The Label

The year 1966 witnessed the publication of Thomas Scheff's *Being Mentally Ill.* Under influences such as Erving Goffman, Edwin Lemert, and Howard Becker,[3] Scheff (1966) argued that being mentally ill was a status conferred on people by others. Scheff's essential individual contribution was to state clearly and organise a sociological viewpoint that had been emerging over the previous few decades. This occurred in relation to such ideas as norms, violations, roles, and labelling—whereby bizarre behaviour may be labelled normal or vice versa (Matza, 1968). In particular, the latter, the long-gestating idea of labelling, now came to dominate critical contemporary thinking.

The year 1967 saw Laing openly declare that schizophrenia was a name for a condition that most psychiatrists ascribed to patients they called schizophrenic (1967a, p. 139). Drawing on Scheff, Laing argued that 'There is no such "condition" as "schizophrenia", but the label is a social fact and the social fact is a political event' (1967b, p. 121). Echoing Gregory Bateson's double-bind hypothesis, behaviour labelled schizophrenic was a strategy to live in an unliveable situation (Laing, 1967b). For Laing, schizophrenic alienation may have had an unrecognised sociobiological function. Schizophrenia was a successful attempt to adapt to pseudosocial realities (ibid). Coincidentally or not, Sartre had written to Laing in 1964, arguing that mental illness was invented to live through an intolerable situation (Burston, 1998). [Following Jung and earlier thinkers, Laing would consequently reconfigure schizophrenia as an inner voyage or metanoia. Beja (1978) also followed the interpretation that schizophrenia was the result of a desperate attempt to avoid insanity. See also Harriman (1947).] In 1960, Laing also identified the schizophrenogenic mother as a possible problem for some patients. But no one had schizophrenia, like having a cold. The patient had not 'got' schizophrenia. The patient's ontological position was that he was schizophrenic (Laing, 1960/1990). For schizophrenia was now a natural healing process that, if facilitated, could enable an existential rebirth. Growth from psychosis was possible (Mosher, 1999). Following similar arguments in *The Divided Self* (1960/1990), Laing further argued for the retention of the name in an existential sense: 'Perhaps we can still retain the now old name, and read into it its etymological meaning: Schiz—"broken"; Phrenos—"soul" or "heart"' (1967b, p. 130).

Laing also asserted that schizophrenia was not a disease entity—but an artefact of capitalist social organisation (Szasz, 2009). This reflected the Marxist thinking that was still vigorous in European society but actively supressed in North America. A similar link between Marx, alienation, and schizophrenia can be found in the work of Joseph Gabel (1960). Gabel (1960, 1997) dismissed the development of schizophrenia as part of a greater history of false consciousness within psychiatry. Similarly, in France, we find the slogan 'Schizophrenics are the proletariat' (Postel and Allen, 1994, p. 387). The politicisation of social and environmental factors did not gain much currency. Not even the Soviets, who maintained their own concept of schizophrenia, would land everything at the feet of capitalism.

Yet, alongside the idea of labelling, a suspicion of social organisation remained in vogue. In 1967, David Cooper also argued that schizophrenia, the existence of which was open to discussion and dispute, was a microsocial crisis. He further argued that 'the process whereby someone becomes a designated schizophrenic involves a subtle, psychological, mythical, mystical, spiritual violence' (Cooper, 1967, p. 13). For Cooper, the disease model was at odds with the very nature of the 'schizophrenic field'. This was the social field in which the label schizophrenia was attached by some participants to others. Hence, like Laing, he, too, argued that schizophrenia was not an entirely meaningless term.

Franco Basaglia, noted for his attempts to restructure the social process surrounding mental illness, accorded a more serious value to the label schizophrenia. By 1968, in Italy, Basaglia, after a reformist period in Goriza asylum, had written his work the *L'istituzione Negata*. And with volunteers from all over Europe he would soon be well on his way to 'depsychiatrisation' (Basaglia, 1968/2013), eventually establishing

Law 180, which, as noted, sought to replace psychiatric hospitals with community services. Following Binswanger, Basaglia thought the schizophrenic comprehensible through existential Daseinanalyse (Basaglia, 1953). The definition of the syndrome [sic] had already assumed the weight of a value judgement, a labelling that went beyond the disease itself. Yet Basaglia acknowledged that diagnosis had the value of a discriminating judgement. Without it, it would be denied that the patient was ill (Basaglia, 1968/2013). In a similar sense, some psychiatrists, as symbolised by Manfred Bleuler, would positively see the term/label schizophrenia as a form of social protection. It protected people from social injustice (Szasz, 2009).

Scheff pushed on. In 'Schizophrenia as Ideology' he now described schizophrenia as an ideology. It was an ideology embedded in the historical and cultural present of the white middle class of Western societies. Scheff further argued that the 'vagueness of the concept of schizophrenia suggests that it may serve as the residue of residues' (1970, p. 17). He thought that those so diagnosed not only explored inner space (Laing's term), but also the normative boundaries of their society. Scheff also argued that ideas such as bizarreness, inappropriate affect, and withdrawal were all dependent on cultural judgements and norms. The label schizophrenia was 'a broad gloss' and the least clearly defined of all psychiatric categories (Scheff, 1970, 1973).

The idea of schizophrenia as a label then existed as a form of outright dismissal. However, in some places it was also thought, at least in a weak sense, to have a residue of utility. This latter idea might even have grown as alternative treatment centres began to verify their procedures scientifically, were it not for one thing. It was about to be blown apart by a social experiment that would rock psychiatry to its foundations.

Rosenhan's Experiment

In 1973, the concepts of schizophrenia and psychiatry were still under attack from critics. To make matters worse, both now also suffered the ignominy of psychologist David Rosenhan's (1973) infamous study 'Being Sane in Insane Places'. Eight pseudopatients (the group included three psychologists and a psychiatrist) had gained access to 12 reputable psychiatric institutions. Often they had simply claimed that they had heard 'unclear' voices, which had merely said 'empty', 'hollow', and 'thud'. In 11 such cases they received the diagnosis of schizophrenia (one case was diagnosed as manic–depressive). Although not recognised as sane by staff (despite quite 'public' shows of sanity), many patients recognised the pseudopatients as sane. When eventually released, after what was for most a disconcerting experience, ten of the 11 received a diagnosis of 'schizophrenia in remission'. The last was released with an unchanged diagnosis of schizophrenia.

Worse was to follow. A teaching hospital, whose staff had heard of the findings before the paper was published, doubted that such an error could occur in their hospital. The experiment was consequently re-run on 193 first-admission patients to the same hospital. The hospital was on high alert for deception. Hence, 41 patients were alleged, with high confidence, to be pseudopatients by at least one member of staff. In fact, Rosenhan had *not sent* any pseudopatients at all.

Rosenhan's results lent credence to a view that he had cited in his introduction. Psychological categorisation of mental illness was useless at best, and downright harmful, misleading, and pejorative at worst. Psychiatric diagnoses were, in the minds of observers, not valid summaries of characteristics displayed by the observed (Rosenhan, 1973). Psychiatric diagnosis, argued Rosenhan, betrayed little about the patients but much about the environment in which an observer found them. The concept of schizophrenia had never seemed so exposed.

In passing, it should be noted that this last part of the study was omitted from a discussion of the Rosenhan experiment by Cooper (2007). This made unsatisfactory an already dubious conclusion that all the research showed is that psychiatrists can be tricked (ibid). Certainly not everyone was entirely troubled. Charles C. Cleland, writing in the *Schizophrenia Bulletin*, thought the study useful in improving diagnostic tools. Cleland facetiously proposed that schizophrenics could perhaps be used to facilitate differential diagnosis. Why? Because it takes one to know one; they had spotted the imposters (Cleland, 1975). However, immediately after the appearance of the study, the *American Journal of Psychiatry* published an article by Ransom J. Arthur, who noted that 'such massive errors certainly cast doubt on the validity of psychiatric diagnosis' (1973, p. 843). Along with other social psychiatric studies, Rosenhan's study had significantly contributed to 'the development of a crisis of identity within the psychiatric profession' (Arthur, 1973, p. 841). Perhaps for such reason we can find that Bellak, denying diagnostic nihilism, would now acknowledge that schizophrenia was merely a nomothetic label. Treatment programmes were aimed at symptom profiles. Agreement on the identification of schizophrenics was a secondary issue (Bellak, 1975).

In 1975, a major *critique* of the study's logic by Robin Spitzer would nonetheless acknowledge that many psychiatrists accepted Rosenhan's thesis. It further admitted that insufficient signs constituted a hitch to a definitive diagnosis of schizophrenia. Rosenhan was entitled to believe that psychiatric diagnoses were of no use and should not have been given to the pseudopatients. Spitzer also noted that many schizophrenia subtypes had not been demonstrated to be distinct subtypes: such subtypes were of no use for prognosis or treatment. The reliability of psychiatric diagnosis in schizophrenia was no better than fair (like many other medical diagnoses). It could be improved (Spitzer, 1975). Similarly, Spitzer would admit a myriad of ways psychiatric labels could be used to hurt rather than help patients (ibid). And that was a critique. By 1978, psychiatrist J.K. Wing, in reference to the study, would continue to speak of elementary errors and psychiatrists using the shakiest of evidence (Wing, 1978).

Adding yet further to this North American crisis was the fact that insurance companies were now grumbling about diagnosis problems. This would potentially deprive the profession of an income source. Central government, mid-recession, was now also reluctant to contribute large funds to psychiatry. That is, unless it became more accountable for its practices (Wilson, 1993). All this took place at a time when controversy raged over the proposed and eventual demedicalisation of homosexuality in 1973. It took place as the Supreme Court was asserting the nondangerous patient/individual's right to liberty (Keith et al., 1976).

The neurologically damaging side effects of some medication were becoming harder to ignore. The drug clozapine was withdrawn in 1975 after it was shown to cause agranulocytosis that led to death in some patients (Gelman, 1999). All this was compounded by the 1975 multi-Oscar-winning adaptation of Ken Kesey's *One Flew Over the Cuckoo's Nest* (rooted in Samuel Beckett's *Murphy*). This led some authors to assert apologetically a 33% decrease in 'Cuckoo's Nest-like state and county hospitals' between 1969 and 1973 (Keith et al., 1976, p. 510).

By 1976, APA president Alan Stone concluded that the profession had been brought to the edge of extinction, largely by social psychiatry (Wilson, 1993). And by now even a troubled *British Journal of Psychiatry* felt compelled to print a heretical article by Szasz. One that declared 'There is, in short, no such thing as schizophrenia' (Szasz, 1976, p. 316). For Szasz, catatonia, hebephrenia, delusions, and hallucinations referred to behaviour not disease. Such terms referred to disapproved conduct, not histopathological change. Schizophrenia was only a word, the name of an alleged disease (Roth, 1977).

Decline

By 1970, a coded article in the *Schizophrenia Bulletin* would speak of the value of antireductionist heretics, who expose weak points in armour and provide a stimulatory role (Dixon, 1970). This seems undeniable. Laing's work, for example, inspired a proliferating network of therapeutic households (Burston, 2001), and other therapeutic centres. Examples included Loren Mosher's 1971 Soteria project, Emanon in 1974, and Crossing Place in 1977 (Mosher, 1999). Nevertheless, although Law 180 had also yet to pass in Italy, signs were already appearing that the tide was turning against a now imploding antipsychiatry and its approaches.

By the early 1970s Kingsley Hall had closed in disarray, with Laing departing for India (Burston, 2001). Other places would continue. However, as Beel would later observe: 'The "angel of love and mercy" career and the Laingian cultural revolutionary who thinks that "crazy is better" because the madman sees to the heart of things—these cannot survive contact with the real world of schizophrenia, and they burn out as they enter its atmosphere' (1975, p. 117). Former patient turned physician Mark Vonnegut (a previous admirer of Laing and Szasz) also condemned Laing in a *Harper's Magazine* article 'Why I Want to Bite R. D. Laing':

> He's said so many nice things about us: we're the only sane members of an insane society, our insights are profound and right on, we're prophetic, courageous explorers of inner space, and so forth. ... But what I felt when I found myself staring out of the little hole in the padded cell was betrayal. I did everything just like you said, and look where I am now, you bastard (1974, p. 90).

Vonnegut's impeccable counterculture credentials (his father was Kurt Vonnegut of *Slaughterhouse 5* fame) made this attack on 'antipsychiatry' as sharp as any psychiatrists themselves had mustered. For some, such protests signalled and accompanied a restoration in the belief and need for a rehabilitated psychiatry (e.g. Mendel, 1976). 'Critical thinking' was also beginning to eat itself. The year 1976 saw the appearance of Szasz's *Schizophrenia: The Sacred Symbol of Psychiatry*.[4] It would argue, along familiar lines, that just like the concepts 'divine' and 'demonic', 'schizophrenic' was wonderfully vague in its content; and terrifyingly awesome in its implications (Szasz, 1976/1988). But the book would now also simultaneously assail Laing and Cooper and the so-called antipsychiatry movement for corrupted radicalism (Flew, 1977).

Also contributing to the decline was the fact that throughout this critical period underlying confidence in a biological basis to mental disorder had held reasonably firm. A breakthrough report of the effects of chlorpromazine by Jean Delay and Pierre Deniker in May 1952 underpinned much of this confidence (Pichot, 1967). The 'cure neuroleptique' was characterised by a disinterested mood, and a relaxation and slowdown of perceptions and reactions (Delay and Deniker, 1955). By 1957, the *BMJ* could report from the second International Congress for Psychiatry, held in Zurich, that 'The tranquillizing drugs were generally agreed to be of value in severe schizophrenic behaviour disorders' (Anon, 1957a, p. 756). Consequently, the introduction of these drugs, and speculations on their mechanisms of action, opened up new lines of biological inquiry and a belief in schizophrenia as a 'chemical imbalance' (Gelman, 1999). And, indeed, by 1970 the *Schizophrenia Bulletin*, while lauding the effects of pharmaceuticals, would also note a corresponding neglect of nonsomatic factors in treatment programs (Anon, 1970c).[5]

Following a 'rejuvenation' of the investigation of the biology of mental health (largely inspired by pharmacology) (Horwitt, 1956), grants for biologically orientated research, as opposed to social research, had remained to the fore (Anon, 1970d). New and promising research vistas, such as Sarnoff Mednick's research on childbirth anoxia, hippocampal damage, and stress continued to emerge (Morriarty and Masset, 1970). Such thinking was reinforced by ever-increasing investigations into biochemical processes, producing notable drug inspired post-hoc theories such as the dopamine hypothesis, which attributed symptoms of schizophrenia to disturbances in dopamine regulation. New statistical approaches were in the air. And computerised tomography, spearheaded by researchers such as Daniel Weinberger, now offered visual evidence of enlarged ventricles and structural abnormalities in the brains of many people diagnosed with schizophrenia (Greenberg, 1979). [But see also pneumo-éncepahlographic research, from 1945, on the third ventrical (Heuyer, 1974).]

The belief in a neurological basis for schizophrenia was further accompanied by at least one firm root in genetic findings and by perceived methodological advances in genetic studies (Rosenthal, 1969). These now offered the possibility of looking for protective or 'antischizophrenia' genes (Gottesman and Shields, 1976). Such thinking was now reflected in PhD oral questions: How might one secure the best odds of selecting a previously undiagnosed schizophrenic individual from the general population, without describing any behavioural trait or symptom? The answer being, 'find an individual *x* who has a schizophrenic twin' (Meehl, 1973, p. 136). If the mental illness called schizophrenia was a myth, argued Seymour Kety, it was a myth with a strong genetic component (Kety, 1974).

Genetic approaches were not flawless. There was increasing scepticism towards Kallman's early twin data (Anon, 1970e). Similarly, there was doubt about the mode of transmission assumptions in the so-called Slater-Böök model, which viewed schizophrenia as largely a single gene defect (Rosenthal, 1977). It was also quickly recognised that fewer than half of the identical twins in studies of schizophrenia had schizophrenia themselves, although they were thought to share all their genes with schizophrenics, therefore underlining environmental factors (Nicol and Gottesman, 1983).[6] Nevertheless, around the 1970s, the genetic school of thought was now seen to be at war with environmental and social schools (Reiss, 1974). The absence of a gene did not rule out genetic evidence (Rosenthal, 1972). And for some optimists, geneticists seemed to be holding all the trump cards (Kessler, 1976).

That said, environmental factors per se could not be ruled out, even if few 'reputable workers' in the USA believed that schizophrenia was now purely a psychogenic disease (Brill et al., 1969). Consensus appeared to stabilise on an old and somewhat familiar middle ground, or for some 'deadlock' (Trotter, 1972). A 1974 literature review would conclude, for example, that 'There is every reason to believe that some biological predisposition—as yet unknown—will be found to interact with social factors—as yet undefined—to cause some fraction—as yet undetermined—of what is now called "schizophrenia"' (Gunderson et al., 1974, p. 49). Similarly, Lidz urged proper consideration of environmental factors—for what he saw as a developmental disorder not a disease (Lidz, 1976). But for many environmental factors now carried the idea of something like an environmentally born toxin or virus. And this could affect the brain much more than, say, social disorganisation within the family.

Finally, alongside new empirical findings, for psychiatrists, the notion of schizophrenia as immeasurably broad could also seemingly now be addressed through the preparation and eventual emergence of DSM-III (see Chapter 9). The architects of DSM-III, convened under an empirically minded Melvin Sabshin, would explicitly push hard to define mental disorder as a medical condition in order to combat antipsychiatry and make psychiatry more scientific (Decker, 2013). [And, by 1983, E. Fuller Torrey would successfully push the idea to a broader public that schizophrenia was a brain disease (Torrey, 1983).] For many of the above reasons, Laing, Cooper, and Szasz, who were believed to have dehumanised schizophrenia by making it a sociological concept, could now be dismissed in the same breath as older psychogenic thinkers like Sullivan and Meyer (Ey et al., 1977). [Ey himself had also wittily remarked that antipsychiatry constituted a return to *'l'anté-psychiatrie'* (Gelinas, 1977).] Similarly, labelling theory could be mocked as absurd (Romano, 1977). And from here on the rhetorical fury of antipsychiatry would fade from mainstream schizophrenia discourse. For many, its companion, the cultural flowering that was the 1960s, already appeared no more than a dead head.[7]

Old Insecurities

Yet beneath the apparent wave of progress, the concept of schizophrenia remained subject to scrutiny in orthodox settings. Old insecurities lurked. As such, when confidence among biologists faltered—as in 1976, when pharmacology studies reported inconsistent findings between laboratories in the arena of platelet monoamine oxidase activity—we find the return of the old nagging and 'serious question': 'What do we mean when we diagnose someone as schizophrenic?' (Wyatt and Murphy, 1976, p. 87). In 1977, the *Schizophrenia Bulletin*, having initiated a review of the clinical picture of schizophrenia, was less confident than ever of what schizophrenia was: 'We had hoped to present a simple statement. Unfortunately, there is not even general agreement as to what schizophrenia is today, no less what it was in 1900' (Durell and Katz, 1977, p. 530). John Ellard, citing genetic advances and defending the changing conception of schizophrenia, would nonetheless admit, 'In fact, we have no general agreement that it exists ... the phenomena of schizophrenia vary quite markedly from culture to culture and from epoch to epoch' (1977, p. 13).

For Stephens (1978) there existed malign and benign forms of schizophrenia. However, whether these forms were discreet entities or points on a continuum could not yet be determined. While Victor Adebimpe, highlighting a case of psychomotor epilepsy misdiagnosed as schizophrenia, noted there was little agreement as to which signs and symptoms were crucial to schizophrenia (Greenberg, 1977).

Although given less attention, wholescale dismissals of the concept continued to emerge. Van Praag, for example, would argue that schizophrenia was an impossible concept. A number of psychoses were being grouped with no points of agreement in respect of symptoms, aetiology, or prognosis. There are hardly any reasons to reduce these disease patterns to a common denominator, either as 'schizophrenia' or as a 'group of schizophrenic psychoses' (van Praag, 1978). Van Praag urged reform of the schizophrenic psychoses.

Conclusion

The vigorous dissent traversing the 1960s represents a critical period in our understanding of opposition to schizophrenia. But perhaps for understanding the historical conceptualisation of schizophrenia this criticism is best viewed as the continuation and manifestation of that which had gone beforehand. It takes its place in a tradition that voiced a ceaseless unease with the concept. For across the twentieth century we have witnessed disparate thinkers who expressed critical attitudes towards the concept of schizophrenia, either in sum or in part.

Nonetheless, for reasons we have discussed, in voicing their criticism, post-1950s thinkers contributed to an atmosphere of renewed scrutiny over what exactly constituted schizophrenia. And the presence of so many critical voices within institutional discourse reflected the fact that schizophrenia was an unstable concept, more than any reflective self-probing by psychiatrists.

Antipsychiatry was declared as being solipsistic and having scandalised, shocked, and disdainfully turned its back on the scientific community. But it was also evaluated in 1977 as acting as a guard dog, of having publically mentioned a certain number of existing confusions and uncertainties in psychiatry. Its value lay in its questions, not its responses (Gelinas, 1977). Nonetheless, the intensity of the criticism of antipsychiatry should not be underestimated. By 1977, one French authority would complain that 'antipsychiatry' had caused schizophrenia to disappear from the garden of nosological spaces (Ey et al., 1977). This, in fact, did not happen, although the statement itself reveals how close the concept's critics seemed to have come to doing so, at least in the minds of their opponents.

Critics of schizophrenia were ultimately dismissed. They hadn't presented persuasive alternatives to schizophrenia (but then an atheist should not have to prove that God does not exist by offering a different God). And the heretical notion of dismissing schizophrenia remained unsuccessful. Although confidence in the concept of schizophrenia itself remained low, schizophrenia was not abandoned. Although at times shaken by criticism, traditional confidence in the biological foundations of mental disorder would never disappear. As we shall see next, a drive was also underway to reform the concept at an institutional level. At no time in its history has the concept ever been thought of as being beyond reform. Yet the DSM-III of the 1980s, in particular, would prove to be one of the most audacious attempts to do so.

9
Manufacturing Consensus in North America

We have already examined how problematic schizophrenia classification, definition, and so forth could be for individual researchers. Yet schizophrenia, as we also saw, would not collapse under the weight of its conceptual problems or critical attention. Indeed, by the close of the twentieth century, schizophrenia had seemingly become a truly international concept, upheld by a global community of researchers. Facilitating all this, and something not yet explored in our earlier chapters, is that over the course of the century, when it came to conceptualising schizophrenia, the once authoritative status of individual researchers was largely displaced by the pronouncements of powerful institutions. Hence, as the twentieth century progressed, attempts to standardise and collectively validate the concept were often made at a communal level through, for example, the auspices of the World Health Organization (WHO).

The importance of the role and contributions of bodies such as the WHO, or regional national authorities, in conceptualising schizophrenia at a communal level cannot easily be dismissed. As we shall see, for example, such bodies facilitated international studies and attempts to harmonise classification. Nevertheless, as the century closed it was increasingly North America, via the APA, that largely controlled the conceptualisation, standardisation, and reification of schizophrenia. And increasingly, and with some degree of success, the APA's DSM would seek to universalise and export to global psychiatrists its vision of the concept. This ascendency occurred even although APA conceptualisation frequently diverged from the vision laid out by the WHO and other international bodies. In its broadest explanation, this remarkable dominance and ambition reflected the growing status and power of North America itself. Such power, which readily translated into research resources, was matched by corresponding weaknesses in the power of its European counterparts and other rivals. For this and other reasons, the APA's understanding of schizophrenia would significantly shape psychiatry's present understanding of schizophrenia. And, as such, a regional understanding of the historical emergence of schizophrenia at an institutional level in North America becomes important when we seek to understand the forces that shape contemporary schizophrenia. A history of the concept is incomplete without such an analysis.

North American conceptualisation ultimately found expression through the APA's DSM series, which focused strongly and at times almost exclusively on taxonomy. Ostensibly, such attempts sought to facilitate better communication among schizophrenia researchers. Yet this institutional emphasis on taxonomy, both here and elsewhere, can be explained as the product of both internal and external forces. For the most part, psychiatry, as we saw, had long possessed an urge to classify and find new species of disorder. However, overlapping and accompanying this urge was the growing institutional desire, as noted by Foucault (1966/2006), to count and quantify the insane and insanity. Such tabulation, evident in North American society since the 1840 census at least, sought—among other things—to generate and standardise figures. It sought to chart the movement of patients, and their financial and economic status. It recorded their perceived race, citizenship, degree of education, marital status, causes of death, and so forth. And, as the twentieth century progressed, such tabulation grew in importance for various public institutions, which, as we shall see, further sought to share and compare this data at an international level.

For North American psychiatry then, as with elsewhere, standardised taxonomy became an increasing bureaucratic and institutional imperative. It was taxonomy, and not symptoms or definitions that appeared most countable. Consequently, the need to produce and collect the results of such tabulation therefore co-existed alongside, and at times transcended, a basic need to facilitate scientific communication and coordinate research efforts. Indeed, as we shall see, at times their tabulation and collection could become an end itself, even if the classes of people that were supposedly being counted were frequently ephemeral abstractions.

In earlier chapters we saw continuous transformation in the concept. We saw frequent disagreement among individual researchers with respect to definition and classification. We also became aware of problems in delimiting schizophrenia's boundaries, of fluctuation in patient's states, and a problem with the presence of other conditions, and so forth. We further saw implicit dissent and actively vigorous dissent, the permeation of the concept by moral concerns, as well as various problems and issues with language itself. Yet this chapter, by contrast, illustrates how institutional conceptualisation of schizophrenia represented a social consensus over and beyond the intellectual concerns and conceptual weaknesses that we witnessed elsewhere in the book.

We will therefore now trace the transformation of schizophrenia's conceptualisation in North America though multiple taxonomic schemes at a communal level. First, we will see an institutional vision of schizophrenia that emerged out of a period of chaos. Second, we will witness its existence in an inter-related system of international and domestic taxonomy, which remained unsatisfactory. Finally, we will examine how a new emphasis on symptom profiles and diagnostic validation gave birth to a schizophrenia that now underpins its modern conceptualisation, as first formulated in the schizophrenia of DSM-III. For all that, a history of most of this period continues to show that even within this display of consensus, schizophrenia remained a deeply unstable concept. Even at an institutional level, researchers simply could not convincingly agree on what it was they considered schizophrenia to be. Instead, as we shall see, what they consistently agreed upon was the continuous need for reform of the concept.

A Statistical Manual

By 1917, the American Medico-Psychological Association (AMPA) and the National Committee for Mental Hygiene (NCMH), funded by the Bureau of Uniform Statistics, produced a plan for the collection of uniform statistics in hospitals for mental disease (which they also promoted to the United States Census Bureau). This resulted in the *Statistical Manual for the Use of Institutions for the Insane* published in 1918 (American Medico-Psychological Association and National Committee for Mental Hygiene, 1918). Prior to the publication of the DSM in 1952, it would become the ostensible de facto standard for statistics and nomenclature in North American institutions for the insane.

In the manual, dementia praecox is given prominence over schizophrenia. The dominance of Kraepelin's concept is unsurprising. Bleuler's ideas were still only entering psychiatric consciousness. In contrast, working translations of Kraepelin's ideas were available in English for over a decade (Kraepelin, 1904/2002; Diefendorf, 1902/1908). The manual simply reflected the comparative dominance of dementia praecox over schizophrenia in North American institutions at this time.

That said, after some anonymous lobbying, the manual also acknowledged that many now preferred the term schizophrenia. This was helped, in part, by growing anti-German sentiment during and in the aftermath of the Great War (Noll, 2011). The manual presented a formulation with four subtypes—paranoid, catatonic, hebephrenic, and simple—that, strictly speaking, conformed to *neither* Kraepelin's nor Bleuler's classification. Bleuler's latent type, for example, was excluded. Similarly, a number of Kraepelin's types were excluded. The existence of doubtful atypical cases was also acknowledged. And the manual noted that cases 'formerly classified as allied to dementia praecox should be placed here rather than in the undiagnosed group' (American Medico-Psychological Association and National Committee for Mental Hygiene, 1918, p. 24). But no actually statistical category was reserved for these atypical cases. We also find a brief reference to autism—a Bleulerian neologism.

Finally, we find the word reaction. This was an impressionistic non-quantative concept of 'habit deterioration'/clear mental disorder or prodromal stages—as promoted by the influential Adolf Meyer. His clinic had been using the term reaction types by August 1906. Similarly, it had been using the term schizophrenic reaction type as early as 1915 (Noll, 2011, 2015)—later interpreted as a reaction to various biological and psychological stresses (Brill et. al., 1969) (n.b. Jung used the term 'reaction types' around 1906, and the '*schizophrenen Reaktionstypus*' appears to exist in European literature from about 1912).

However, despite the apparent credentials of this particular manual, it was unsuccessful. From 1918 to 1952, no one authority succeeded in homogenising psychiatric classification and imposing it on individual researchers and their institutions. We have already seen some of the challenges concerning classification that would exist in such an undertaking. However, much of the blame, it has also been argued, can be placed at the feet of Meyer, who held extraordinary influence in the USA between 1902 and 1940. Meyer largely avoided quantitative analysis. He further repudiated the demand for disease specificity in psychiatry and was reluctant to create a common scientific language about the insanities (Noll, 2011). Indeed, the Meyerian school was held to have an almost pathological fear of diagnostic labels (Skottowe, 1940).

As such, in addition to progressions of the *Statistical Manual for the use of Institutions for the Insane*, multiple other taxonomic schemas also co-existed and evolved to produce new headaches. The American War Department, for example, in their *Outline of Neuropsychiatry in Aviation Medicine*, classified schizophrenia into catatonia (two forms), paranoid, hebephrenia, and simple forms—which, following Kahlbaum, they called 'heboidophrenia'. The latter gave rise to the 'heboidophrenic', that is, many a 'criminal, hobo, prostitute, crank, and eccentric' (War Department, 1940, p. 93) ('heboid' had actually been rejected by Bleuler in 1911). Next, there was a 'mixed form' that was 'very common' 61 (War Department, 1940, p. 95). Finally, there existed a sixth form for the unrecorded or *unclassified* cases. In such cases individuals—even when encountered by the psychiatrist—were 'without an aggregate of manifestations permitting definite classification in any one of the different forms' (War Department, 1940, p. 96). They appear *not* to have displayed significant quantity of symptoms to find themselves classified as suffering a mixed form of schizophrenia. This held even though they supposedly indelibly bore 'the stamp of the disease' (ibid).

By 1949, Philip Ash could alarmingly demonstrate that in 52 white males, psychiatrists could only agree on a diagnosis of schizophrenia 20% of the time (Decker, 2013). Multiple classifications clearly hadn't helped. The 1952 introduction of DSM-I described the previous period of classification of disorders as *chaotic*. It described one agency as having used one nomenclature system for clinical use. The same agency had another for disability ratings, and an international classification for statistical use. [The most important international classification would be the series known as the *International Classification of Diseases* (ICD), in existence since 1910.] DSM-I further stated that *every* teaching centre had also made its own modifications to existing systems of classification (American Psychiatric Association, 1952).

Nevertheless, for bureaucratic administrative purposes and research quantification, these schemas possessed a primitive but important utility value, irrespective of their validity and intercompatibility. Hence, seemingly useful, if uncertain, figures could be generated that estimated schizophrenia struck 1–2% of the population of the USA (Anon, 1947). In general, major revisions only occurred every decade or so. But although official classification rarely kept up with the latest theoretical ideas regarding schizophrenia/dementia praecox, its official and communal nature meant that they increasingly acquired authority status. Twentieth-century advances in the fields of statistics and calculation greatly added to the allure of such status. And where used, their descriptions became normative and sanctioned, all of which, to a great extent, reified an institutionalised interpretation of dementia praecox, schizophrenia, and their subtypes. By mid-century, however, the presence of acute logical discrepancies, predicated upon the presence of strong social forces, was becoming more and more difficult to hide.

ICD-6

In 1948, ICD-6, of the aforementioned ICD series, was published (*Manual of the International Statistical Classification of Diseases, Injuries and Causes of Death*). Its authors were international experts on health statistics, working under the auspices of the WHO. A vast number of other organisations were involved in its formulation, including the United Nations Statistical Office and the International Labour Organization. And, as a series, the ICD had found widespread usage outside of North America.

The ICD now takes on a particular importance in American psychiatric history. As DSM-I later noted, international treaties signed by politicians had ordered that future national classifications *should be in agreement* with this international classification (American Psychiatric Association, 1952). The political agreement in question here is reflected in the WHO 'nomenclature regulations 1948' detailed in ICD-6. Essentially, scientists would have to agree on their systems of knowledge because politicians had decreed they ought to (see also Zajicek (2014) on how political censure in 1932 by the Commissariat of Public Health helped shaped the Soviet conception of schizophrenia). As will be made clear, this policy eventually contributed to disruption, distortion, and, in addition, to a modicum of farce in the classification of schizophrenia.

ICD-6 was something of a hodgepodge in itself, laden with shortcomings. In its defence, the manual's introduction quoted one Professor Major Greenwood: 'The scientific purist, who will wait for medical statistics until they are nosologically exact, is no wiser that Horace's rustic waiting for the river to flow away' (World Health Organization, 1948, p. xiv). Under the heading 'Schizophrenic Disorders (Dementia Praecox)', itself placed under the heading 'Mental, Psychoneurotic and Personality Disorders', it seemingly relegated dementia praecox to secondary status. This was in line with its fifth incarnation from 1940, which had merely listed 'schizophrenia (dementia praecox)' (American Psychiatric Association, 1968). Yet seemingly copying Bleuler's *Textbook of Psychiatry*, it now detailed two parallel classifications: one for dementia praecox, the other for schizophrenia.

Supposedly unifying the two vertical lists, we find that each successive subtype of dementia praecox is matched with its conceptual equivalent in schizophrenia. Each instance is then united with a numbered code. Hence, for code 300.3 both types have a 'Paranoid' type. On face value, this was similar to what Bleuler himself had earlier advocated.

Hence, all seemed okay. But inevitably such a strategy ran into problems when attempts at one-to-one mapping were made. Kraepelin and Bleuler had detailed different numbers of subtypes.

There was, for example, no dementia praecox equivalent for code 300.4: 'Acute schizophrenic reaction'. Nor was there one for Code 300.6 'Schizo-affective psychosis', which included three forms: 'Mixed schizophrenic and manic-depressive psychosis', 'Schizo-affective psychoses' and 'Schizothymia'. The latter was supposed to be a *personality disposition*. However, it didn't fit into the manual's listing of pathological personality disorders that elsewhere included the schizoid personality (code 320.0). Similarly, for code 300.2 'Catatonic type' dementia praecox allowed for two forms of catatonia 'Catatonia' and 'Dementia, catatonic'. This contrasted with the singular 'Schizophrenia, catatonic'. And with code 300.5 we simply find 'Latent schizophrenia', which subsumes 'Latent schizophrenic reaction' and 'Schizophrenic residual state (*Restzustand*)'.[1]

As such, although the manual seemed to imply synonymy between dementia praecox and schizophrenia, it simultaneously—and accurately—revealed this not to be the case. The section concluded with code 300.7, 'Other and unspecified'. In this code we find 'schizophrenic reaction', and dementia praecox under 'Dementia praecox NOS' (not otherwise specified) or 'any type not classifiable' under the earlier listed codes. Clearly, leaving room for *any type not classifiable* had something for everyone. As noted, it was widely adopted for *official* use outside the USA (although not necessarily loved by researchers).

DSM

By 1952 the APA introduced the latest incarnation of its official statistical manual. The new manual, entitled *The Diagnostic and Statistical Manual for Mental Disorders*, aka DSM-I, marks the beginning of the modern era of psychiatric classification.[2] As noted, under international treaty, the USA had effectively committed the DSM to agreeing with the ICD, which still drew attention to dementia praecox. However, in DSM-I, the term dementia praecox was downgraded (although still declared a synonym). It now seemingly *gave way* to nine 'schizophrenic reaction types': simple, hebephrenic, catatonic, paranoid, acute undifferentiated, schizo-affective, childhood, and residual (American Psychiatric Association, 1952, p. 5). The schizophrenic reaction types were found under 'Disorders of psychogenic origin or without clearly defined tangible cause or structural change'.

Schizophrenic reactions were grouped under psychotic disorders. They were characterised by fundamental disturbances in reality, regressive trends, bizarre behaviour, disturbances in stream of thought, and/or by formation of delusions and hallucinations. The schizophrenic types, the reader was informed, had been increased in number and type to allow more detailed diagnosis. Childhood type was introduced for those children who before puberty would display psychotic reactions manifesting primarily as autism. Chronic undifferentiated, the manual noted, also included 'latent', 'incipient', and 'pre-psychotic' reactions. It also included anything else showing more than a schizoid personality but which remained *unclassifiable*.

The manual next attempted an appendix of cross-coding to allow the manual to be compatible with the ICD series. This was 'an effort of no small note' (American Psychiatric Association, 1952, p. xi). To be precise, it failed. A number of issues emerged. ICD, to give a simple example, had no classification for DSM's childhood type. But rather than cross-tabulate the DSM with ICD code 300.7, 'Not otherwise specified', the DSM authors equated it with code 300.8 in the ICD manual. Such a code, however, *did not exist* in the ICD manual. Indeed, the fictional code did not exist in the next ICD revised edition, which occurs in 1957 (World Health Organization, 1957). [We also find that in its appendix, the DSM additionally reintroduced the catch-all subtype 'Other and unspecified' (but *not* in the manual proper). This it equated with ICD code 300.7, 'NOS', which it further recoded as 'Other and unspecified except childhood type.']

Finally, the DSM manual authors admitted in its appendix that they had adjusted the international classification (unilaterally) and that people using such codes should note the discrepancy. Unsurprisingly, then, the manual also admitted in the introduction to its appendix that *incompatibilities* between the two manuals exist. Later in 1959, the then DSM-II committee chairman, Moses M. Frohlich, would acknowledge that the APA had disappointed many in the WHO. It did so by adopting a 'quite different classification in 1952 than the one our representatives urged on the World Health Organisation in 1948' (Zubin, 1961, p. 87).

ICD-7 and ICD-8

No great breakthrough was made with the 1957 publication of ICD-7. It was largely unchanged from ICD-6. And in some countries it was distinctly unpopular. In England and Wales, for example, although the Ministry of Health had requested this nomenclature to be used across

Great Britain, the request was widely ignored. Not to be deterred, the Ministry of Health statisticians set about *translating* everything into ICD-7 format. They did so in order to fulfil their *international obligations*. The rules were somewhat arbitrary. A diagnosis of 'probable schizophrenia', for example, would be coded as 'Schizophrenia unspecified', whereas a diagnosis of 'suspected schizophrenia' could be coded as 'Diagnosis uncertain'. This resulted in distortion and confusion in perceived rates of schizophrenia (Cooper et al., 1972). ICD-8 would fare little better in 1968. Within a few years of its publication, Aubrey Lewis would describe it as a hotchpotch that flies in the face of taxonomic rectitude. However, for Lewis, it persisted 'for lack of anything better, which would be generally acceptable' (1979, p. 193). One thing ICD revisions did do, however, was, in light of international treaty obligations, to prompt and excuse further revision of North American classification.

DSM-II

Many American psychoanalysts, as typified by Karl Menninger in 1963, still preached against an over-reliance on classification and diagnosis (Decker, 2013). This disposition ultimately saw them cede control, to their own detriment, of important positions on committees relating to this ostensible backwater. Yet the 1968 DSM-II probably did little to dissuade them of their viewpoint. Schizophrenia was now simultaneously referred to as schizophrenia (singular), and the schizophrenias, a group of disorders (plural). Out of almost nowhere it now included 13 categories: simple, hebephrenic, catatonic (withdrawn and excited), paranoid, acute schizophrenic episode, latent, residual, schizo-affective (excited and depressed), childhood, undifferentiated, and other (and unspecified) types. Without any obvious sense of irony, Cantor in the same year (1968) would suggest the term 'occult schizophrenia' for all ill-defined schizophrenia.

The manual notes that several new subtypes of schizophrenia had been added by subdividing old categories. For example, catatonia had been divided into withdrawn and excited. Schizo-affective had been divided into excited and depressed. In addition, the latent type had been added for disorders previously labelled unofficially as incipient, prepsychotic, pseudoneurotic, pseudopsychopathic, or borderline schizophrenia. The latter followed 1939 descriptions by Adolph Stern (Decker, 2013), and was also sometimes referenced as compensated schizophrenia (Spittell, 1979). The quixotic circumstances leading to this rampant subdivision were best summarised by Chairperson E.M. Gruenberg in the manual's preface. Gruenberg noted, 'Even if it had tried, the Committee could not establish agreement about what this disorder is; it could only agree on what to call it' (American Psychiatric Association, 1968, p. ix).

Earlier, in 1961, DSM-II chair Frohlich had noted that the nascent DSM-II 'represented a compromise between the various private and semiprivate classifications previously evolved and used in the various teaching centres, hospitals, state mental health systems, armed services and so on ...' (Zubin, 1961, p. 88). To Frohlich, this was philosophically possible because the differences in various classifications seemed 'to be more semantic that real' (ibid). Driving Frohlich's philosophical flexibility was doubtless a general tendency in the APA, as with all professional bodies, to seek influence over areas that threatened, or which it deemed constituted, its domain of expertise. As such, while Frohlich agreed that classification needed to be simple, he reasoned that it must also be 'flexible and expandable so that it can include or be compatible with all kinds of detailed refinements and private classifications without too much difficulty' (ibid). Inevitably, the criteria by which things were divided into classes could often only be established 'by common agreement, or convention, or arbitrary decision' (Zubin, 1961, p. 87).

By 1968, the classification of schizophrenia in the ICD and DSM remained incompatible. DSM-II acknowledged it was unable to reconcile the many views of its members. And, in contrast to ICD-8, it continued to assert the existence of a 'schizophrenia childhood type'. Hence, it noted that 'This category is for use in the United States and does not appear in ICD-8' (American Psychiatric Association, 1968, p. 35). Technically, this was legally possible because under WHO regulations an escape clause, 'exceptional circumstances', allowed such manipulation. A *modified* ICD version of ICD-8 'adapted for use in the United States' then came into being. It *included* schizophrenia childhood type. It also placed under other forms of schizophrenia the category *infantile autism* (US Department of Health, Education and Welfare, 1972).

DSM-II now abandoned the use of the term 'reactions'. The manual denied that this signalled 'a return to a Kraepelinian way of thinking, which views mental disorders as fixed disease entities' (American Psychiatric Association, 1968, p. 122). The denial seemed plausible because the equally out of favour term dementia praecox was now also completely abandoned (its absence went without comment). That said, if one did want to use the term dementia praecox then it could be found in the ICD-8 manual adapted for use in the USA.

It was classified under schizophrenia code 285.9 'Unspecified type' as 'Dementia praecox not otherwise specified' (US Department of Health, Education and Welfare, 1972).

Outside of official documentation new subtypes had also continued to emerge. For example, in 1968, the year of the manual's appearance, a discontented Kline would give us the following triad: (1) schizophrenia—childhood asocial type, (2) schizophrenia—fearful paranoid, and (3) schizophrenia—schizoaffective (Stephens, 1978). This did not necessarily mean the manual could simply be discarded. Such formulations never gained enough support to displace official taxonomy. And, in at least some locations, use of the manual was mandated by insurance companies (Greenberg, 1977).

Post-1950, the concept of schizophrenia had been engulfed in a period of heated criticism, some of which we have dealt with previously. DSM-II had done nothing to alter that situation. As the 1960s closed, Loren Mosher (1969) declared that progress in schizophrenia had also been impeded by fragmentation. There existed conflict and isolation of schools within the fields of research, training, and treatment. Mosher was then leader of the Center for Studies of Schizophrenia at the National Institute for Mental Health (NIMH). As late as 1977, Romano would characterise the previous two decades of diagnosis and classification as possessing 'a certain looseness, if not slovenliness' (1977, p. 544). Yet behind the scenes things had been slowly beginning to change. And schizophrenia research now entered a period in which a number of important overlapping events and studies occurred that would shape the future of the concept.

The period witnessed new thinking on what constituted important symptoms. And it saw a renewal of interest in statistics and the methodology surrounding diagnostic criteria. Notably, two international studies, 'The Cross-National Project' and 'The International Pilot Study', also examined the reliability of the concept. Results were not impressive but they foreshadowed the kinds of investigation that would be witnessed in the eventual construction of the APA's hegemonic DSM-III. All of this casts great light on the concept of schizophrenia post-1960 in the immediate period leading up to DSM-III. Let us briefly examine each of these studies in turn.

Whichophrenia

By 1970, Charles G. Costello would argue, in *Symptoms of Psychopathology*, that present classifications systems were inadequate, primarily because they were premature. Classes had been formed by *majority vote* of psychiatric professional *associations* rather than by scientific investigations. The continued reliance on traditional diagnostic categories, he reasoned, would appear to be 'quite unwarranted and might be a disservice to both the psychiatric discipline and the psychiatric patient' (Costello, 1970, p. 2).

For Costello, what was important was the application to schizophrenia research of yet more new statistical systems, such as multivariate statistical analysis. In 1971, psychiatrist Heinze E. Lehman similarly saw the now exploding field of statistics as a way to,

> save psychiatric nosology from the confusion created by such rapidly developing new 'syndromes as schizomania, phasophrenia, schizonoia' (Marchais, 1966), eliminate the uncertainty of 'whichophrenia' (Altschule, 1967), [and to] correct the American tendency to over-diagnose schizophrenia (a tendency to which British psychiatrists have referred as schizophrenomania) (1971, p. 143–144).

Such enthusiasm partially stemmed from the increasing availability of the computer, the status its use conferred, and its possibilities. Consequently, accompanying evidence for seven new subtypes of schizophrenia, Gerard (1964) had earlier included a photo of Mr Nils Mattson, a project statistician, with an all-authoritative computer printout. 'These programs have the capacity to make psychiatric diagnoses that are 100 percent reliable', boasted one set of researchers (Strauss et al., 1974a, p. 41). Such aspirational thinking was further accompanied by ever more powerful statistical techniques. In 1977, for example, Bartko and Carpenter used cluster analysis to discover the four subtypes: 'flagrant schizophrenia', 'insightful schizophrenia', 'typical schizophrenia', and 'hypochondrical schizophrenia' (Bartko and Carpenter, 1977). (Although quixotically, their statistical data were further transformed into four different *faces*. Each was supposedly representative of a type of schizophrenia—and work was in progress to refine the model, with depictions of hair, cheeks and tongues (ibid).)

Statistical and indeed methodological optimism contrasted favourably with the predominance of the case study over quantification in early twentieth-century research (although never exclusively so). It also represented an advance over pseudoequations. For example, when, J.S. Beck (1954) managed to contrive an admittedly apologetic equation for schizophrenia: S = d/er (schizophrenia = demands on the person over ego resistance). Similarly, statistical optimism represented quite a change from earlier comments by authorities such as Manfred Bleuler. In 1951, he had thought it utterly impossible to investigate the course of schizophrenia (for all forms) on the basis of valid statistical procedures (Stephens, 1978). [This statistical turn was not something that came out of thin air. In many senses it was the culmination of events rooted in the late nineteenth century, out of which a growing acceptance of disease specificity, as well as the need for quantification and laboratory work, slowly articulated itself (Noll, 2015).]

The Cross-National Project

At the earlier mentioned 1959 conference 'Field Studies in the Mental Disorders', where definition had caused so much friction, debate had also centred on classification. Psychiatrist H.C. Rümke, for example, proposed that one of the contributions of nosology was the clinical differentiation of the group of 'genuine schizophrenia' from cases of 'pseudoschizophrenia'. The latter could be differentiated further into five subcategories: (1) endogenous pseudoschizophrenia; (2) exogenous toxic schizophrenia; (3) characterogenic pseudoschizophrenia; (4) developmental pseudoschizophrenia; (5) cerebro-organic pseudoschizophrenia. To this he added (6) cases that cannot be properly classified. Admitting nosology was progressing slowly, Rümke nonetheless defended the use of nosology, and argued against seeing nosology as 'the pursuit of a phantom' (Zubin, 1961, p. 77) (the expression itself can be found much earlier in the 1912 writings of Alfred Hoche in criticism directed at Kraepelin's classification (Hoche, 1912; Noll, 2011)).

Rümke's new classification, however, was not even in accord with the people in the room. Joachim-Ernst Meyer, for example, alternatively proposed that for international purposes, schizophrenia should be divided into a broad and simple scheme, namely schizophrenia, schizophrenic episodes, and paranoid states. For conference organisers psychologist Joseph Zubin and psychiatrist Paul Hoch it now became clear that 'conferences would not settle the matter and that a field study was required' (Cooper et al., 1972, p. ix). As such, they set about organising the 'United States–United Kingdom Diagnostic Project' (also titled 'The Cross-National Project for the Study of the Diagnosis of Mental Disorders in the United States and the United Kingdom').

The Cross-National Project would analyse the reliability of psychiatric diagnosis across a sample of patients, from 22 London hospitals and nine New York hospitals. It began in 1965. In doing so, disorders that the interim DSM-II would deem *separate* to schizophrenia were contained within the term. These included paranoid states, paranoia, and involutional paraphrenia (Anon, 1974a). Similar research elsewhere had been ominous. In 1961, Biostatistician Morton Kramer had found that the rate of admissions into UK hospitals for manic–depressive disorders was ten times that of the USA. And differences in schizophrenia diagnosis were detectable, too (Kramer, 1961).

One finding was that, in general, American psychiatrists diagnosed schizophrenia in a wider variety of clinical conditions than their British or British-trained colleagues. New York psychiatrists tended to diagnose schizophrenia more readily than their London counterparts. This occurred no matter what their age, country of origin, length of training, type of practice, theoretical orientation, or academic status. Psychopathology generally regarded as *characteristic* of schizophrenia might be reported as *present* in a patient by American psychiatrists but as *absent* by British psychiatrists (in the same patient). Descriptive psychopathology was shown to be subject to distortion by systematic bias.

Although the findings were 'disconcerting', the study also noted transatlantic diagnostic agreement could be obtained on *some* patients. This led to the hope that perhaps a core group of patients with schizophrenia could be identified. For the team this possibility provided a 'partial validation of the concept of schizophrenia' (Klerman, 1989, p. 28). Both narrow and broad concepts of schizophrenia had theoretical advantages and disadvantages. But until one could be proved better than the other, the study suggested that for now,

> assignments could be made to categories composed of cases that 1) both British and American psychiatrists would call schizophrenia (i.e., 'agreed schizophrenics'), 2) both would call something other than schizophrenia (i.e., 'agreed nonschizophrenics'), and 3) only the Americans would call schizophrenic (i.e., 'broad concept schizophrenics'). Such labels, while not flouting conventional diagnostic usage, would acknowledge some of the forces that lead to cross-national miscommunications (Anon, 1974a, p. 99).

Few researchers responded to this proposal to flout conventional diagnostic usage. Partial validation was seemingly not enough. And indeed for other critics, diagnosis was now dismissed as a sacred cow (Gunderson et al., 1974). Yet for some psychiatrists it was better than nothing.

First-rank Symptoms

Complementing a growing interest in assessing the reliability of classification was a renewal of interest in symptoms. Notably, the 1950s saw the first English translation of German Karl Schneider's *Clinical Psychopathology*. In order to facilitate diagnosis, Schneider had listed a number of first-rank symptoms. These would soon gain international recognition:

> Audible thoughts, voices heard arguing, voices heard commenting on one's actions; the experience of influences playing on the body (somatic passivity experiences); thought withdrawal and other interferences with thought; diffusion of thought; delusional perception and all feelings, impulses (drives), and volitional acts that are experienced by the patient as the work or influence of others (Schneider, 1959, pp. 133–4).

Schneider further noted that when *any* of these modes of experience were 'undeniably present and no basic somatic illness can be found, we may make the decisive clinical diagnosis of schizophrenia' (1959, p. 134). Furthermore, 'Symptoms of first-rank importance do not always have to be present for a diagnosis to be made' (1959, p.135). And that 'we are often forced to base our diagnosis on the symptoms of second rank importance, occasionally and exceptionally on mere disorders of expression alone ...' (ibid).

Up until 1961, references to Schneider's work in the Anglophone psychiatric literature were fairly rare. Nevertheless, in an overview of *Contemporary European Psychiatry*, an enthusiastic 1961 review put Schneider—then retired—firmly on the map. It claimed that Schneider wanted to establish a biological psychiatry. And it hailed Schneider 'as the logical successor to Kraepelin' (Hoff and Arnold, 1961, p. 62) (remarks in 1951 assessing German Psychiatry by Henri Ey may also have proved influential (Ey, 1996)). Only later would claims be made in some quarters that about 42% of the time *none* of the first-rank symptoms were present (Strauss et al., 1974a). And, indeed, some critics complained that Schneider's concepts had become so hallowed that they essentially had become fossilised in some circles. No merit had been given to subsequent attempts at improvement (Stephens, 1978).

Schneider's symptoms received yet more attention when the WHO study described Schneider's work as a possibly fruitful approach to defining schizophrenia in purely symptomatological terms (World Health Organization, 1975). Simultaneously, Bleuler's fundamental symptoms were compared *unfavourably* to those of Schneider (ibid).[3] The WHO study did not actually use Schneider's symptoms. Instead, it relied on a check list of 360 questions from an assessment tool known as the Present State Exam. Nevertheless, by 1978, one of the WHO document's principal authors, J.K. Wing, in contrast to earlier scepticism, further declared that 'about two-thirds of all patients given a clinical diagnosis of schizophrenia described experiences equivalent to Schneider's first rank symptoms. ... This confirmation ... makes it worthwhile to examine them in detail' (1978, p. 5). Schneider had not actually been able to indicate one pathognomic symptom. And sometimes he relied on mere expression to make a diagnosis. Yet with the aid of this kind of endorsement, Schneider's first-rank symptoms came to be viewed as canonical symptoms worthy of consideration in schizophrenia conceptualisation. And, ultimately, as we shall see, they would find partial expression in DSM-III.

The International Pilot Study

A year after the 1974 UK–US study, the International Pilot Study on Schizophrenia produced its results. This study involved over 150 staff from numerous disciplines. Notables included Zubin, J.K. Wing, T.Y. Lin, N. Sartorius, and Morten Kramer. Some 1202 patients, 811 diagnosed with schizophrenia, were seen in nine countries. These included Taiwan, Columbia, Czechoslovakia, Denmark, Nigeria, the UK, the USA, India, and the USSR (World Health Organization, 1975). The study gave training to psychiatrists and used ICD subclassification.

In spite of attempts to standardise the methodology, not all locations would conform. Moscow, for example, used periodic schizophrenia, sluggish schizophrenia, chronic undifferentiated schizophrenia, and Melekhov's shift-like schizophrenia, or *schub*, from the German word attack (which had to be cross-tabulated into ICD). Most of these seemingly had numerous subcategories (Brown et al., 1974). Furthermore, some centres possessed broader interpretations of the concept than others.

Moreover, the results were alarmingly inconsistent. Moscow and Washington could diagnose no catatonic cases between them (despite having the two broadest conceptualisations of schizophrenia in the study), while Agra (India) could diagnose 22 (out of 54 found in all centres). Similarly, Moscow and Washington could only diagnose one hebephrenic, in contrast to 22 being found in Columbia's Cali (World Health Organization, 1975). Hence, in relation to subtypes in particular, the study highlighted clear conceptual problems. Unsurprisingly, by now the *British Medical Journal* had already reported that subtyping had fallen into relative disuse (Anon, 1974b).

For some critics, the absence of perfected psychobiological or physiological indicators meant that psychiatry was deemed to be 'inextricably trapped in circular reasoning' (Hogarty, 1977, p. 588). Where consistency was found, all that had been shown was that psychiatrists could be trained to be 'reliable' on certain scales (Hogarty, 1977). Yet the International Pilot Study was hopeful that it had contributed to the *beginning* of a process of identifying a core group of people with schizophrenia (World Health Organization, 1975). As with the Cross-National Project, it was believed such studies could be improved upon.

Feighner's Criteria

In parallel with the ongoing Cross-National Project and the International Pilot Study, North America now saw the emergence of the St Louis Diagnostic Criteria. These diagnostic criteria were developed at the Department of Psychiatry at Washington University, St Louis, Missouri. Summarised in a highly cited 1972 paper, the criteria became commonly referred as 'Feighner's criteria' after its principal author, John P. Feighner. The Feighner criteria aimed to provide firm rules. These would enable a decision to be made in every case, on whether or not the criteria for a diagnosis of schizophrenia were satisfied. The system used a 'menu' format that would foreshadow the 'operational' diagnostic criteria for schizophrenia in DSM-III. It allowed for just three subtypes: paranoia, catatonia, and hebephrenia. The authors favourably contrasted all this against DSM-II. Their diagnostic classification had been based upon the best clinical judgement and the experience of a committee and its consultants. The authors of the Feighner paper claimed it had been validated by follow-up studies. They further provided a framework for comparison of data gathered in different centres that served to promote communication between investigators (although even supporters acknowledged such tests required the utmost fastidiousness) (Feighner et al., 1972). As it happens, Feighner's criteria also built on a literature base that was not always empirical, as Feighner committee member Rodrigo Munoz would later acknowledge. This disclosure was at odds with claims to synthesis based on data rather than opinion on tradition (Decker, 2013).

Feighner vs. Schneider

By 1972, Robert Kendell sought a remedy for diagnostic confusion in schizophrenia. He argued that a choice should now be made between the Feighner criteria or Schneider's criteria (Kendell, 1972/1975). The problem with Schneider's criteria, thought Kendell, was that between 20% and 30% of patients who would probably be regarded as schizophrenic by most psychiatrists did *not* possess them. This figure actually varies in the literature; for Harrow and Silverstein (1977), for example, 49% of patients did not have any first-rank symptoms. In any case, Kendell believed that the problem with Feighner was that it included 'thought disorder'. Thought disorder, he believed, had defied all attempts to have its essence captured (Kendell, 1972/1975).

Schneider's first-rank symptoms were also problematic for others. One 1973 study, by Carpenter and Strauss, 'did not support the common European view that schizophrenia is a qualitatively distinct diagnostic entity with pathognomic signs and symptoms and predictable course' (Gunderson et al., 1974, p. 21). Elsewhere, Martin Roth thought that Schneider's symptoms may not have been able to delineate nuclear schizophrenia but they had at least facilitated the identification of these syndromes in the 'penumbra' around schizophrenia. Diagnosis could not be based on Schneider's first-rank symptoms alone. A *wider* range of criteria had to be drawn upon. Roth (1978) had in mind *thought disorder*, which, in contrast to Kendall, he believed to be an ambiguous phenomenon only in its milder form. A debate concerning thought disorder therefore lurked in the ether while Schneider's symptoms were increasingly questioned (perhaps at times overstating Scheider's actual claims).

Many moved towards adopting Feighner. Kendall, for example, now also pushing the operational agenda, appears to have given the nod to the Feigner criteria; it possessed modest but useful longitudinal criteria. Others also asserted that the Feighner criteria constituted an important exception to the belief that North American schizophrenia was broader and less well defined (Strauss et al., 1974b).

The Feighner criteria, in turn, were 'improved upon' (in terms of inter-rater reliability) by one of its original authors, Eli Robins, working with psychiatrist Robert Spitzer and psychologist Jean Endicott. Endicott and Spitzer had previously worked on evaluating computer-aided diagnosis on the United States–United Kingdom Diagnostic Project, and psychopathology scales. [The Fortran-based computer program Diagno I was based on DSM-I and was, unsurprisingly, criticised for incompatibility with ICD-8 (Cooper et al., 1972). Another program, Catego, failed to consider patient history.] All this resulted in a set of criteria known as the Research Diagnostic Criteria (RDC) (Spitzer et al., 1975). The RDC outlined 25 diagnoses as opposed to Feighner's 16 or DSM-III's ultimate 265. And it would be principally this modified version of the Feighner criteria that DSM-III would eventually use (yet, as we shall see, with a restricted reliance on Schneider). [The RDC also echoed an earlier 1962 brief psychiatric rating scale, propagated by mandatory use in NIMH drug trials in 1968 (Overall, 1979; Overall and Gorham, 1962).] Broader social factors were also at play. For example, Roger Blashfield complained that Feighner's criteria became well known, at least in part for social reasons. They had been promoted and cited by the 'invisible college' that was the prolific St Louis/Iowa group (Blashfield, 1982).

The Neo-Kraepelinian Revival

There certainly were strong social groups driving the conceptualisation of schizophrenia in North America at this time. And individual movers within such groups, and their effects on the production of DSM-III, have since been well documented by Decker (2013). One of the most powerful individuals in question was Robert Spitzer. He had consulted on DSM-II and had been head of DSM-III since the project's inception in 1974. Possessing psychoanalytic training he seemed like a perfect candidate to an APA leadership still dominated by psychoanalytic leanings. However, Spitzer had turned from his roots. Once in control, the dynamic and driven Spitzer mainly appointed people who were firmly set against psychoanalysis and acted accordingly (Decker, 2013). Psychoanalysts, largely antipathetic to classification, were caught sleeping. By the time they were alert to the danger they were caught largely flat-footed in their responses. Spitzer had earlier demolished Lauretta Bender's finding of a 'primitive reflex' in schizophrenic children by showing the reflex existed in all children. He now thought redefining schizophrenia was one of the biggest challenges of DSM-III (Decker, 2013).

In contrast to prior DSM committees, Spitzer was not content to assess the existing scientific evidence in order to make decisions concerning schizophrenia. He thought there were many 'different conceptions' of schizophrenia (Greenberg, 1977, p. 29). Instead, where Spitzer felt evidence was lacking (e.g. schizophrenia's diagnostic reliability) he set about organising studies for DSM-III. Symptom course, hypothesised prognostic features, and descriptive 'axes' for items such as social functioning were now in vogue. Spitzer was accompanied in his actions by those of other significant actors such as Eli Robins, one of the principal authors of the Feighner criteria.

But collectively on a broader social level, the various protagonists now shaping mainstream schizophrenia conceptualisation came to be known as the neo-Kraepelinians. Ten neo-Kraepelinians would make up the 19-strong empirically minded psychiatric task force that created DSM-III (Blashfield, 1982). These would validate the diagnostic instruments and methods they trusted most. Where doubt remained, they would create new instruments and methods. Yet, irrespective of evidence gathering, in 1989, Gerald L. Klerman would acknowledge that 'DSM-III was developed by consensus rather than by reference to an existing body of empirical knowledge' (1989, p. 30).

The neo-Kraepelinian position first gained significant attention having been outlined as such by Klerman in a 1978 manifesto or 'credo'. Klerman declared that American, British, and Canadian psychiatry was then in the midst of a Kraepelinian revival. The term neo-Kraepelin can be found as early as 1939, in the *Journal of Criminal Psychopathology*. Hence, even before DSM-III and Klerman's manifesto, Howard Goldman (1977) (pushing a nonunitary disease hypothesis) had suggested that the DSM's neo-Kraepelinian nosology had not been that useful. Klerman, however, traced the origins of the revival to the textbook *Clinical Psychiatry* by Mayer-Gross, Slater, and Roth. This textbook had first appeared in the 1950s. And it drew attention to promising genetic research (Mayer-Gross et al., 1955).

The neo-Kraepelinian manifesto was largely a combination of idealism, platitudes, and rhetoric. It variously attacked approaches found in psychoanalysis and countered ideas from heretics such as Szasz. But, importantly, it essentially affirmed a belief in discreet mental illnesses—the biological roots of which should be methodically investigated (Klerman, 1978, pp. 106–7). For Klerman, schizophrenia was best regarded as an illness, and the disease concept was the most applicable way of describing it. If schizophrenia was a myth, noted Klerman, then the individuals who were schizophrenic were doubly delusional in their suffering. With further investigation psychiatry would probably reaffirm Bleuler's concept. However, admitted Klerman:

> Before one can conclude definitively that schizophrenia is a disease, conclusive evidence will have to be presented as to etiology and clinical course ... such evidence ... does not yet exist for schizophrenia (1978, p. 111).

Indeed, by 1989, Klerman would still continue to note that the assertion that schizophrenia was a disease 'remains only a belief until appropriate evidence is gathered' (1989, p. 29).

DSM-III

Schizophrenia now found itself in a period of active reconceptualisation. In just six years prior to the publication of DSM-III, the taskforce issued a large number of communications relaying 'progress' and new ideas. An enormous raft of publications supporting taskforce opinion, critiquing rival systems, making pronouncements, and providing occasional refutations of critiques were issued leading up to DSM-III. In total, using successive drafts of the DSM-III, 12,667 patients had been evaluated by 550 clinicians in 212 facilities. New tools such as the schedule for affective disorders and schizophrenia were invented. Questionnaires were given to clinicians and feedback was elicited. DSM-III claimed far greater reliability than DSM-II. Although meetings had often been fractious, the taskforce had liaised with many organisations. These included the American Academy of Psychoanalysis, the American Psychoanalytic Association, and the American Psychological Association. Where differences had been left unresolved, the manual would later claim, they had at least been clarified (American Psychiatric Association, 1980). The proposals of many other schizophrenia theorists faded in comparison. Although it contained 'inaccuracies', thought critical psychiatrist Alan Taylor, it would be more rigorous and restrictive than DSM-II, 'the current diagnostic bible' (Anon, 1978b, p. 231).

With a new emphasis on operational procedures, reliability for schizophrenia diagnosis would eventually be boasted as much improved (k = .82 when two clinicians interviewed a patient together). And this achievement would remain the outstanding feature of DSM-III's reconceptualisation of schizophrenia. However, despite all the taskforce work to secure advances in reliability, many of the decisions made in reconceptualising schizophrenia appear to have been arbitrary.

A new category, 'Schizoaffective disorder', for example, was 'a compromise between two extremes: those who consider it a subtype of Schizophrenia and those who consider it a form of Affective Disorder' (Spitzer et al., 1978, p. 491). Indeed, the official summary of a 1976 progress conference acknowledged that classification systems had to be a product of many compromises. Just over half of field participants surveyed thought schizoaffective merited its own category. About one-quarter of the participants remained unsure. It was bounced from category to category. It was a form of schizophrenia in an April 1977 draft. Ultimately, following William Carpenter's thinking, it was relegated to the amorphous 'Psychotic disorders not classified elsewhere'. The task force, noted Spitzer, gave up attempting to provide specific guidelines.

Studies showed its reliability to be at best fair. Even less agreement was found for 'Confusional schizophrenia'. It was removed from early drafts in 1976. Disputes extended to fundamental ideas about schizophrenia. In one November 1976 conference, a clash occurred between Paul Wender and Roy Grinker over the existence of psychological aetiology in schizophrenia. Grinker, advocating for such an aetiology, remarked that at times being in the meeting was like being in never-never land (Decker, 2013).

The removal of simple schizophrenia had been a personal objective for Spitzer. Spitzer had further objected to borderline schizophrenia. This was because he thought there was no such thing as 'mild' schizophrenia (Decker, 2013). And ultimately Spitzer's position won out. The diagnosis of schizophrenia would be restricted to those who had been overtly psychotic (delusions, hallucinations, or grossly disorganised speech) (American Psychiatric Association, 1980). This meant that latent, simple, or borderline schizophrenia cases would now be moved to 'schizotypal personality disorder' (formerly known as schizoid). As such, 'simple schizophrenia', once thought by Bleuler to be a subtype of schizophrenia, was now a personality disorder.

Objections to the removal of simple schizophrenia or the shifting of latent and borderline to personality disorder, such as those made by John Racy and Richard Ciccone of Rochester school of Medicine, were seemingly rejected (Greenberg, 1977). Similarly, Roy Grinker (1979) unsuccessfully argued that borderline syndrome represented an independent entity, and that a newly proposed schizotypal category was unsatisfactory. Perhaps for this reason, however, the manual would strike a compromise with the note that schizotypal personality disorder could still be present in 'Schizophrenia, residual type' (American Psychiatric Association, 1980).

The schizophrenia of DSM-II was largely broken up. One important communication noted that the concept of schizophrenia in DSM-II had now been subdivided into a number of different categories. Many individuals who formally would have been diagnosed as having schizophrenia in DSM-II would probably now be diagnosed according to DSM-III as having something else, for example 'Paranoid Disorder, Schizoaffective Disorder, an Affective Disorder, Schizophreniform Disorder, Brief Reactive Psychosis, Atypical Psychosis, or Schizotypal Personality Disorder' (Spitzer et al., 1978, p. 489). Nancy Andreason, a key DSM-III taskforce member, later nonchalantly explained how some of the conceptual change had actually come about. She noted, 'With a few swift strokes of the nosological scalpel, many of these forms of traditional schizophrenia were dissected away and included in other categories' (Andreason, 1994a, pp. 354–6).

Other background forces also lurked around DSM-III's formulation, although their precise role in relation to schizophrenia is less clear. Comments by Guze, for example, revealed that DSM-III decisions makers were deeply concerned about the possibility of giving insurance companies an excuse not to pay up. Hence, they sometimes *modified* their opinions as a result. Various other aspects of the DSM-III draft were also proving highly controversial, possibly draining attention away from schizophrenia. External pressure also came from the need for compatibility with ICD-9. And, indeed, such was the contentiousness of the manual overall that on one occasion DSM-III was almost replaced by ICD-9 (Decker, 2013). Simultaneously, time pressures also forced decision making.

Aware of the impending arrival of DSM-III many urged caution. Psychiatrist John Romano argued that diagnosis, classification, and treatment approaches were prone to being influenced by *fads* in choosing diagnostic subtypes of schizophrenia. In addition, there existed a tendency to resort to chronic undifferentiated schizophrenic subtype as a diagnostic expedient (Romano, 1977). Romano applauded the insistence on greater specificity and explicitness of rules in the DSM-III draft. But he wondered if the search for reliability might lead to a loss of validity. Others asserted that the haste to construct diagnostic instruments could overlook family history and long-term outcome as means of defining schizophrenia (Vaillant, 1978a). While yet others cautioned that 'symptoms could be the worst features for biological researchers to anchor on, since they may be primarily determined by interpersonal psychosocial factors' (Buchsbaum and Haier, 1978, p. 474).

The APA accepted a final draft of the manual in May 1979. By 1980, DSM-III's operational vision of schizophrenia officially arrived. It included the subtypes 'disorganised', 'catatonic', 'paranoid', 'undifferentiated', and 'residual' (somewhat more than the three envisioned in the Feighner criteria). The manual notes that it has excluded illnesses without overt psychotic features. This included latent, borderline, and simple schizophrenia. It stated that they were more likely to be diagnosed within the manual's section on personality disorders (e.g. schizotypal personality disorder), but that the clinician wishing to use these non-DSM-III diagnoses could do so—as they were included in ICD-9 (American Psychiatric Association, 1980).

At an institutional level then, schizophrenia, for all its improved reliability, remained schizophrenia à la carte. This was the kind of classic fudge only a committee could think of or accept. ICD-9 (published in 1977) had provided ten subtypes of schizophrenia. These included simple, hebephrenic, catatonic, paranoid, acute schizophrenic episode, latent schizophrenia, residual schizophrenia, schizoaffective type, other specified types of schizophrenia, and unspecified schizophrenia. All sorts of contradictions arose. For example, the schizoaffective was *excluded* from DSM-III schizophrenia, and made a separate disorder, but was present in ICD's schizophrenia.

The Schneider question also remained unresolved. Commenting on the history of the DSM-III process, Nancy Andreason noted that early drafts of DSM-III had initially included Schneiderian symptoms. However, 'a series of reports appeared in the literature indicating that Schneiderian first-rank symptoms were not pathognomic of schizophrenia' (Andreason, 1994b, p. 345). Because of such reports, 'Schneiderian symptoms were de-emphasized in DSM-III' (ibid). Yet, in fact, Spitzer and colleagues had hedged their bets and concluded, 'that given our current knowledge (ignorance?), some of Schneider's first rank symptoms are useful in the diagnostic criteria' (Spitzer et al., 1978, p. 492, parenthetic expression in the original). Hence, although Schneider was seemingly de-emphasised, Anderson admitted (correctly) that in the concept's description 'The description of characteristic symptoms placed great emphasis on Schneiderian first-rank symptoms' (Andreason, 1994a, p. 357). In other words, DSM-III possessed a description of schizophrenia that *emphasised* Schneiderian symptoms. Yet it also included a checklist of criteria that *de-emphasised* Schneiderian symptoms.

As noted earlier, the dissociation or spaltung that Bleuler deemed *central* to his conception of schizophrenia had been entirely extracted from the concept of schizophrenia. It was now in another category: 'Dissociative disorders' (Garrabé, 2003). Schizophrenia was also conceptually narrowed in one instance because a concerted campaign to see post-traumatic stress disorder included in the manual had defeated arguments from Washington University psychiatrists. The latter had thought war veterans should be diagnosed under existing categories, such as schizophrenia (plus alcoholism and depression) (Decker, 2013). And, indeed, many of the numerous new DSM-III classifications probably also impacted upon the concept in ways that have yet to be fully assessed.

DSM-III Attacked

DSM-III was warmly received by many in mainstream psychiatry as a paradigm shift whose operational methodology signalled progress. Nonetheless, DSM-III schizophrenia was immediately attacked by a vocal minority. This foreshadowed extensive critiques of the entire concept of schizophrenia by Mary Boyle (1990) and Richard Bentall (2003).[4] In 1981, for example, Wayne S. Fenton, Loren, R. Mosher, and Susan M. Mathews reviewed six systems for diagnosing schizophrenia, *including* the nascent DSM-III. None of these systems, the authors argued, had established construct validity. They were all, in a sense, arbitrary. Furthermore:

> the elevation of any one diagnostic system to an official status is thought to be premature, clinicians and researchers alike are advised to exercise caution and open-mindedness in their use of DSM-III. There is as yet no evidence that its criteria for schizophrenia are either less arbitrary or better (in identifying a group of 'true' schizophrenics) than those of other systems or DSM-II (Fenton et al., 1981, p. 452).

For psychologist Paul Meehl, the construct validity of the whole class of schizophrenics had been increased. He did not think that most researchers would find it too much effort to use DSM-III in their work. Nevertheless, he critically noted how DSM had eliminated some signs and symptoms that some clinicians had been *relying* on. These included symptoms such as autism and ambivalence—'considered fundamental by the master himself' (Meehl, 1986, p. 219). Meehl further complained that it had removed an instance close to his heart: 'anhedonia'.

Moreover, it was incontrovertibly possible, argued Meehl, that some patients sharing the underlying aetiology and psychopathology of the core group of schizophrenias would now *not* be detected by DSM-III criteria. In the context of discovery, the way clinicians categorise their world would determine what they were capable of noticing. Hence, 'When one disperses a group of people who are heterogeneous in some respect, but homogeneous in some core feature of high causal relevance, into a number of heterogeneous diagnostic categories the best bet is that they will get lost in the shuffle' (Meehl, 1986, p. 220).

Meehl further found the attitude of unnamed dogmatists, or 'vulgar operationalists' malignant. In doing so he spoke of a 'chilling effect'. This was a subtle kind of social process, whereby research proposals were being *rejected* on the grounds that they did not employ official categories. Meehl added, 'I have heard research-orientated clinicians express concern about this ... and one sometimes hears it alleged that it has occurred' (1986, p. 217). Meehl stressed that pressure should not be put on researchers. Nor should they be punished financially or otherwise, for delineating further conjectures, entities, or dimensions of their own.

Hans Eysenck first called the manual 'absurdly detailed', which after reading his final comments reads like a compliment. DSM-III had superficially provided some five axes in order to further the characterisation of schizophrenia, against the preference of some DSM architects such as Eli Robins (Decker, 2013). Yet Eysenck observed nonetheless that,

> It is interesting to note that the question itself (categorical or dimensional) is hardly ever seriously asked by psychiatrists, and that they show little interest in biometric methods designed to answer it, or the results of such studies (1986, p. 78).

DSM-III, to Eysenck, was at best little other than subjective ratings and descriptions strung together without any quantitative elaboration.

> This is not the way of science, and the fact that large committees have decreed that DSM-III represents an optimum description of psychiatric reality does not alter the fact that it is based on a fallacy, namely, the fallacy of categorical differences between groups (1986, p. 91).

Eysenck consequently proposed a direct move away from a categorical diagnostic system altogether. Instead, Eysenck favoured a dimensional approach, in which the 'psychiatric universe' would be structured along three major dimensions: psychoticism, neuroticism, and extraversion–introversion. A dimensional system, argued Eysenck, 'would not insist on diagnoses useful mainly for administrative rather than medical or scientific purposes' (ibid). For Eysenck, DSM-III exemplified an anti-scientific and irrational approach to perfection. It laid down laws as if no rational person could quarrel with it:

> The fact that such an approach and such an empty, atheoretical, and antiexperimental system can find acceptance in psychiatry say more about the nature of modern psychiatry than any critic, however hostile, might be able to say (ibid).

Drawing on the inconsistence in the aforementioned study by Fenton et al. (1981), Eysenck concluded:

> It is necessary to throw out the whole approach, hook, line and sinker, before anything better can take its place. DSM-IV, if ever such

> a misshapen fetus should experience a live birth, can only make confusion worse confounded and make the psychiatric approach to classification even less scientific than it is at the moment (1986, p. 96).

Psychologists Esther D. Rothblum, Laura J. Solomon, and George W. Albee now also began to question the genetic basis of schizophrenia. If the diagnosis of schizophrenia was unreliable, they argued, a similar warning applied to genetic studies (Rothblum et al., 1986).

Slowly, even supporters of DSM now began to register their problems. J.K. Wing, for example, thought that 'schizophrenia is a mental disorder (or group of disorders)' (1985, p. 1219). He admired DSM-III. Nevertheless, he still criticised it for rules that could not be validated and for leaving much room for variability. Five years after the publication of DSM-III, Wing noted that 'how many symptoms and in what combinations are required for a diagnosis remains a matter of opinion' (ibid).

DSM-III-TR

The DSM-III manual proved highly lucrative for the APA. Within another seven years, the APA brought out the innocuously sounding DSM-III-R. Ostensibly, 'R' stood for 'Text Revision' (American Psychiatric Association, 1987). On close inspection, it made fairly substantive changes to the concept of schizophrenia, contrary to popular belief both then and now. DSM-III, for example, had included the provision that onset must be before the age of 45 years. However, the text revision quietly dropped that requirement (ibid). Few supported what had, in fact, been little other than an arbitrary cut-off point included by the 1980 committee. This had rendered anyone on or after their forty-fifth birthday incapable of developing and suffering from schizophrenia. This restriction had theoretically removed late schizophrenia (*spätschizophrenien*) described by Bleuler, which supposedly occurred in the fifth and six decades of life. And, similarly, it had affected such concepts as Fish's 'senile schizophrenia' and the *altersschizophrenien* (old-age schizophrenia), described by Janzarik, which supposedly occurred in the seventh and eight decades of life (Fish, 1960).

Also, in the revised edition we find that section 'A' of the diagnostic criteria, which mandated at least one of six criteria during a phase of an illness, was arbitrarily reorganised 'to make it simpler'. It now divided the six into three sections, one of which had to be present for at least a week, unless the symptoms were successfully treated. Section 'B' was further modified to let the phrase 'childhood onset' slip back in, all reference to childhood having been omitted from DSM-III.

In addition, as the DSM-IV taskforce would later admit, DSM-III-R retained the same subtype categories, but 'significantly altered the criteria for the paranoid, disorganised, and undifferentiated subtypes' (McGlashan and Fenton, 1994, p. 419). Finally, swinging away from Schneider once again, the descriptive listing of four Schneiderian symptoms was reduced to two, while more emphasis was given to negative symptoms 'by increasing the relative weighting of flattened and inappropriate affect' (Andreason, 1994a, p. 357).

The net effect of both DSM-III and DSM-III-R, thought Nancy Andreason was 'to reduce the boundaries of the concept of schizophrenia to a relatively narrow construct and to require the presence of psychotic symptoms for a diagnosis of schizophrenia' (1994b, p. 345). DSM-III-R was supposed to be only a text revision, with no major changes. However, mused Andreason a little later, DSM-III-R 'appears to be even narrower than DSM-III' (1994a, p. 374). After the publication of DSM-III-R, Thomas McGlashan declared the Heisenberg Uncertainty Principle [sic] was at work in schizophrenia: 'the entity you are measuring moves simply by virtue of how you define it' (1988, p. 533).

Conclusion

The twentieth century saw institutionalised attempts to pin down schizophrenia classification in North America. The conflicting research of individuals would now be downplayed in favour of achieving a broad social concept of schizophrenia. Imposing a sometimes fudged social consensus on madness was not a by-product of twentieth-century scientific discourse. It was now integral to it. Following Kraepelin's early strategy, institutionalised approaches lived in the hope that such taxonomy could pave the way for the elucidation of biological mechanisms in the future.

Yet each new committee, rejecting the work of predecessors, changed the concept in various ways. Among compromise and disagreement, subtypes were subtracted or added. The concept's boundaries ebbed and flowed. Although claims were made to the contrary, this often meant that the DSM was incompatible with other institutional attempts to conceptualise schizophrenia, most notably ICD. Yet political agreements had legislated against such a possibility. The results were sometimes farcical. However, the process gained a modicum of transparency. DSM-III saw greater reliability in diagnosing schizophrenia than had hitherto been demonstrated. Yet decisions were not always based on empirical evidence. As the 1980s ended, the concept of schizophrenia remained highly contentious and in flux.

McGlashan's objections took place against an industry of schizophrenia research, which now marched on unrepentant. This research was now boasting new statistical technologies and advances in areas such as gene sequencing methodologies. And accompanied by an as yet unrealised faith in an ultimate teleology for the classification of schizophrenia it remained undeterred. Arguably, DSM-IV and DSM-IV-TR would trundle out yet more of the same. Such taxonomy could occur, even if this meant persevering with a façade of classification that even the defenders of classification appeared to have their doubts over.

An alternative investigative approach to characterising schizophrenia used polydiagnostic studies. Such studies could use multiple forms of classification. This could include alternative or multiple versions of the DSM and ICD or synopsised Bleulerian criteria, in addition to many other new classification schemes, most notably the St Louis Diagnostic Criteria. In 2006, Lennart B. Jansson and Joseph Parnas would evaluate 92 such studies:

> The polydiagnostic studies do not provide sufficient validity data to justify claiming a clear superiority of any particular definition over others. In many studies, the percentage of sz [sic] cases so diagnosed by all diagnostic algorithms is remarkably low (2006, p. 1178).

Jansson and Parnas (2006) would report that this was not reflective of a class with a particularly strong validity and what was conspicuously lacking in the polydiagnostic studies was a serious and systematic reflection on the conceptual validity of schizophrenia, that is, what one took this illness to be in the first place. This work emerged from an earlier study that examined eight separate systems, including ICD-9, ICD-10, RDC, and DSM-IV. After excluding the ICD-10 simplex category they concluded that 'there are only 14 cases diagnosed as schizophrenic by all 8 systems and 108 patients diagnosed as schizophrenic by at least one of the systems' (Jansson et al., 2002, p. 111). Jansson and Parnas also echoed earlier comments by Holzman and Matthysee, who, in a review of schizophrenia genetics, argued that reliance on classification left investigators open to error. For Holzman and Matthysee, revised diagnostic systems such as DSM-II, DSM-III, and DSM-III-R, and several versions of the ICD, suggested that 'it may be argued that the same deck of symptom-cards has only been periodically reshuffled and redealt' (1990, p. 281). [They urged reform through expansion and refinement of the phenotype (i.e. schizophrenia), including a more prominent use of psychological methods.]

Conclusions: Twentieth-century Schizophrenia

By examining the historical representation of schizophrenia through symptoms, metaphor, definitions, and classification we have broadly surveyed schizophrenia's conceptualisation. Behind the façade of heroic psychiatrists, we have witnessed the formation of a complex social concept. We have begun to sketch what might be called an *epistemological sketch* of twentieth-century schizophrenia. In doing so, we have furthered our understanding of a concept that an iconoclastic Szasz (1976) once referenced as the greatest scientific scandal of our scientific age.

Schizophrenia's contradictions and histories are many. Collectively, researchers variously affirmed numerous abstractions called schizophrenia. Each variant was assumed to somehow reference a stable, invariant, transhistorical object, or objects, of ontological inquiry. All this eventually gave way to an institutionalised and tightly controlled operational understanding of madness. It may not have appeared quite as 'obscenely interpretable' as hysteria did (Micale, 1995), but schizophrenia in its totality was historically contingent. It was insecure and shifting. It was perpetually rebeginning. At times it lacked epistemological rigour, integrity, and overarching coherence. It embraced myth, forgetting, disappearance, and transformation. Indeed, it was, to paraphrase Foucault, much more historical than is usually believed.

The Split Personality

In the twentieth century, there were many stigmatising attitudes towards those diagnosed with schizophrenia. These often implicated wider society. In 1974, for example, the *Schizophrenia Bulletin* would discuss the 'negative halo' of stigma. This stemmed from the belief (then fading) that mental illness was a punishment from God (Rabkin, 1974).

Nevertheless, we now understand that one particular stigmatising force—the split personality—cannot be dismissed as a popular misconception. Schizophrenia, we have discovered, was negatively associated with the split personality—and to a lesser extent Jekyll and Hyde—by members of the behavioural professions. As such, people diagnosed with schizophrenia were haunted by the concept's metaphorical content; it imputed nefarious properties that did not exist (at its nadir, the poster for the 1976 slasher movie *Schizo* read 'Schizophrenia, when the left hand doesn't know what the right hand is doing'). This fits with much of what we know about the stigmatising and punitive function of metaphor in other diseases (Micale, 1995). Nevertheless, twentieth-century psychiatry did not abandon usage of the name. It had forgotten Bleuler's emphasis on splitting and its own role in what it called a public misconception. Instead, over time, schizophrenia partially came to be characterised in opposition to this metaphor.

Symptoms

Schizophrenia's symptom profile changed throughout the twentieth century. Some symptoms such as *Benommenheit* have largely disappeared from the literature. This occurred even though Kraepelin had noted in 1893 that *Benommenheit* was present at the beginning of dementia praecox. And even although Bleuler had speculated it might be a *primary symptom*. The importance of yet other symptoms, such as autism or blocking, which was once thought by Bleuler pathognomic of schizophrenia, has been severely downgraded. Some symptoms appear to have existed only as isolated singularities, reported by only a single or small number of observers. Still other symptoms were critically scorned, as when Sullivan referred to those of the 'side-show variety'. And groups of symptoms, for example those outlined by Schneider, or those collated as catatonia, have passed in and out of prominence. One medical student, voluntarily hospitalised with a diagnosis of schizophrenia gave a more alarming account. The student noted that with *all experiences* being ineluctably reinterpreted as *symptoms*, it produced the feeling of being subhuman (Anon, 1977a).

As historical emphasis on given symptoms has changed, we also looked at myth-making and forgetting in our understanding of historical symptoms. The 'Four As', we discovered, were rooted in salient neologisms such as autism and ambivalence. Each 'A' was not without merit. In particular, loosening of associations and disturbances of affect played a central role in Bleuler's thinking. However, as an object of investigation, taken as a collective representation of Bleuler's thought, the 'Four As' mnemonic was a distortion and simplification of Bleuler's thought. And, just as with the case of splitting and the split personality, myth-making and forgetting entangled themselves in the conceptualisation of schizophrenia.

Definition

Through definition, we examined the historic lack of agreement and inability to find an essential characterisation of schizophrenia. And ultimately in North America we witnessed the side-lining of definition for the operational approach of the APA's DSM. Prior to this, the masking of one label by another (i.e. schizophrenia over the 'obsolete' dementia praecox) had superficially created an illusion of progress. Psychiatry and psychology appeared to have advanced their knowledge base. In fact, behind the name, there existed a multitude of conflicting theoretical positions. The results were not trivial. For example, Soviet definitions of schizophrenia excluded information that would be crucial to an American psychiatrist. Hence, for Shapley, 'The two cultures each of which has its own concept of mental illness in general and schizophrenia in particular, disagree on who to call schizophrenia' (1974, p. 935). The very multiplicity of definitions seemed to point to their collective insufficiency and a more profound weakness within the concept itself.

The Morally Insane

Those who sought to conceptualise madness dragged into their net of classification and definition the marginalised, the vulnerable, and the dispossessed of twentieth-century society. Such characters make for interesting reading. They included vagabonds, transvestites, homosexuals, bisexuals, hoboes, sexual perverts, exhibitionists, and fetishists. They included eccentrics, lounge lizards, old maids, army deserters, conscientious objectors, political dissidents, and book readers. Other groups touched were Jews, prostitutes, criminals, happy hooligans, bobby soxers, nagging housewives, those unable to work, cranks, litigious individuals, and pseudogeniuses (McNally, 2009).

All this was not so much novel. Rather, it constituted part of an occasional reaffirmation of society's need for moral breaches to be explained and, where possible, to have them allocated into a scientific category such as dementia praecox or schizophrenia. Sakel's moral insanity was in harmony with this tradition. As is well known, for historical reasons, psychiatry partially emerged out of a penal mentalité. And it never made a clean break with it (Foucault, 1961/2006). Accordingly, it is unsurprising in one sense that schizophrenia, in its twentieth-century conceptualisation, singled out for attention the morally insane. Indeed, it would have been *more surprising* if such a tradition was *absent* from the history of schizophrenia. For, as Richards has stated, 'a society's concept of madness is necessarily also a statement of its concept of normality' (2002, p. 192).

Classification

In classification, a botanic tradition underlay the endless search for new species of schizophrenia. Throughout the book, we have mapped a good portion of the competing schizophrenia classifications and subtypes scattered through the twentieth-century literature (although doubtless not all). Their continuous emergence bears some resemblance to a comment made by Ludwik Fleck in 1935. Fleck noted, 'The explanation given to any relation can survive and develop within a given society only if this explanation is stylized in conformity with the prevailing thought style' (1935/1979, p.2). So when a paper's title now asks, 'Do Cenesthesias and Body Image Aberration Characterize a Subgroup in Schizophrenia?' (an echo of Huber's 1957 cenesthopathic schizophrenia) (Röhricht and Priebe, 2002), we now see such a question with fresh eyes. Such a theorised subtype competes with many others. Without social support, it may well remain localised and never find support at institutional or international level. And there is a strong possibility that it merely represents a somewhat arbitrary theoretical abstraction, stylised in the dominant mode of psychiatric expression, rather than something of utility in clinical practice.

In parallel with a century of mutating official taxonomy, local and regional classifications proliferated throughout North America and elsewhere. As such, the literature is laden with complaints such as, 'European psychiatrists, especially those in Scandinavia, France, and Germany, have for many years criticised Americans for having too wide a concept of schizophrenia' (Stephens, 1978, p. 26), and 'there is no uniformity of diagnosis within the United States or Europe' (Stephens, 1978, p. 25). Similarly, the National Institute for Mental Health's David Rosenthal would note of the 'reactive form of schizophrenia' that 'The Europeans say it is not schizophrenia at all' (Trotter, 1972, p. 59) (Rosenthal thought Europeans preferred to call it psychogenic psychosis or schizophreniform psychosis). Such differences should not be underestimated. In 1969, an experiment involving a filmed psychiatric interview of a young woman in her 20s took place. It revealed that one-third of the American psychiatrists involved made a diagnosis of schizophrenia. Yet none of the British psychiatrists put this forward as the primary diagnosis (Katz et al., 1969).

To the historical observer the restless mutation in classification was characterised not so much by striking progress, but by appearance and disappearance. The moment we study the concept in one place and time, certain forms of schizophrenia appear to vanish or appear doubtful in many others. In possessing this quality of polymorphism, schizophrenia revealed itself as many things, including conditions such as encephalitis. Moreover, we further saw how certain subtypes were used to describe the truly ethereal; for example, kinds 'in which the characterising classificatory types have dropped out of the picture' (Hoskins, 1946, p. 91). Similarly, the literature boasted kinds that were considered *unclassifiable, not otherwise specified,* or *yet to be discovered.*

Small wonder then that Szasz called schizophrenia a *panchreston,* an 'explain-all' term explaining everything and nothing. Discussing schizophrenia in a similar manner, the report on *Psychiatric Diagnosis in New York and London* could speculate that 'every classification has to have, in practice if not in theory, at least one category which is only loosely defined and can act as a "rag bag" for patients who do not fit in elsewhere' (Cooper et al., 1972, p. 129). Hence, this procrustean concept simultaneously appears to have revealed itself as nothing. All of this calls to mind an insightful observation by Richards (2002): not everything which can be measured necessarily exists.

Understandably, we now have little appetite to retroject the concept's polymorphous nosology prior to 1908, the year of its formation. We can see the concept's historical limitations. Even without historically contextualising prior centuries, we are less convinced than ever that Socrates suffered schizophrenia (Zilboorg, 1942), or that, 'Countless schizophrenics, judged to be possessed by the devil, were burned at the stake ...' (Lehman, 1980b, p. 1105). Indeed, following Foucault, it clearly seems that this concept cannot presuppose the conditions of its own possibility.

Foucault's response to a devastating caricature of classification by Jorges Luis Borges was to laugh uneasily. However, this French intellectual was not alone in laughing. At the conference 'Problems in Field Studies in the Mental Disorders' things had also descended into tragic comedy. One participant shared the following joke: 'Let's call all cases schizophrenia, and then try and find out what's wrong with the patient' (Zubin, 1961, p. 117). At the same conference, a schizophrenia researcher, Alexander H. Leighton, also provoked laughter as he joked:

> The term social psychiatry gives more pause. I suspect the discrepancies between us as to what we would mean by this term would be even greater and wider than those which would arise if we had taken a vote on what we mean by thermonuclear schizophrenia (Zubin, 1961, p. 324).

Similarly, one Lilienfeld amused with: 'Psychiatrists do disagree, and they disagree not only with respect to diagnoses, but they disagree as to whether or not they agree [laughter], and this is quite confusing to a non-psychiatrist' (Zubin, 1961, p. 208). Much later, the pendulum swung to narrower definitions of schizophrenia. Then, Gerard Hogarty would quip that 'It is amusing to fantasize that if the move toward narrowness continues, schizophrenia might well disappear by the 21st century' (1977, p. 591).

We should not dismiss these jokes lightly. For, as with Foucault, at the source of this laughter lay unease. For Foucault, this unease led to intense philosophical and historical rumination on madness and *unreason*. Yet for twentieth-century psychiatry the concept's supporters remained stubbornly defensive in the face of its self-contradictions. And in the words of Pierre Bourdieu (1990), each generation seemingly naturalised its own arbitrariness.

Contesting Schizophrenia

Twentieth-century schizophrenia was more highly contested than first meets the eye. Criticism was not confined to narrow quarters, external observers, or the period that was characterised by a heretical and contentious 'antipsychiatry'. Across the century we have seen how key researchers questioned the concept's definition and its imprecise boundaries; how they used disparaging terms like trough, dump pile, wastebasket diagnosis, catch-all, imprecise nosological fiction, and residue of residues. Hence, as much as any antipsychiatrist, it was researchers like the biologically inclined Hoskins (1946) who suggested schizophrenia might be a semantic convention. This was a research community that was frequently divided and at odds with itself. There was habitual dissatisfaction and calling into question of the concept.

In spite of a crisis of conceptualisation, criticism of schizophrenia and its classification was largely ineffectual. Indeed, against the thousands of research findings produced on schizophrenia and its subtypes every year, such criticism would fade into the margins, its most radical protests ignored. The concept survived particularly well at an institutional level where, for bodies like the World Health Organization and the APA, refinement was always deemed possible. Nevertheless, criticism existed here, too. And communal revisions frequently occurred in a climate of widespread disagreement. As such, the incessant transformation in the official conceptualisation of twentieth-century schizophrenia does not read as a narrative of linear progress or conceptual clarification. Rather, at least in one sense, schizophrenia's narrative arc, or constant mutation, is better understood as a continuous critical rejection of earlier conceptual formations.

Caveats

Although we have neglected them, it also goes without saying that to understand this twentieth-century concept, the personal histories of Eugen Bleuler and other historic figures were not without importance. Who could read A.A. Brill's following observation and not suspect that Bleuler's personal circumstance affected his formulation of schizophrenia?

> When I was in Zurich, Bleuler used to tell us that we could influence even the worst catatonics by suggestion. He gave his own sister as an example. She lived in his home in the hospital, and from my room across the hall I could see her walking to and fro monotonously all day long. Bleuler's children ... when they wanted to climb anywhere, they would use her as though she were an inanimate object, like a chair. She emanated no affect, and the children had no affective relationship with her. Bleuler once had occasion to move her when she was in an acute state of excitement. He did not want to use force, and he thought he would try suggestion. He told us that he worked on her hour after hour, talking to her and urging her, and at last she dressed and went along with him. Bleuler cited this as evidence that you *can* do it (1949, p. 28).

The fact that Bleuler's own sister was seemingly catatonic is genuinely fascinating. And it may have profoundly influenced his thinking. Speculatively, Bleuler may have rooted his therapeutic optimism in familial necessity. Equally, it is possible that he also feared that he, too, might potentially have succumbed to his own disorder. This was erroneous as it happens, although at least one obituary labelled him eccentric: this was a class of people he declared to belong to schizophrenia and which others such as Phillipe Khouri once described as 'near-schizophrenia'. Did a fear of a personal lurking madness explain why his formulation of schizophrenia—notably latent schizophrenia—seemed to edge at times towards 'normality'? This is an open question, although, unfortunately, Bleuler's letters sent to Freud for analysis no longer appear to exist.

There are other interesting nuggets, such as the fact that Bleuler supported eugenics (1911/1952, p. 473), or that in 1930 he charged F. Scott Fitzgerald an exorbitant $500 fee for less than a day's consultation with respect to Zelda Fitzgerald—who thought Bleuler an imbecile (Cline, 2002). Moreover, the case books, stories, and biographies of patients are loaded with materials that are anything but mere nuggets. Some are known, some not. Collectively, they constitute a rich source of understanding schizophrenia that must ultimately be interpreted through broader sociocultural analysis. I have already mentioned some of this book's many other limitations in its introduction and outline, and the reader should tease them out appropriately. Doubtless there are more. But, fortunately, the history of schizophrenia is a collective endeavour and many of these limitations are addressed elsewhere in some shape or other. Truth be told, it is probably better to speak of 'the histories of the schizophrenias' rather that the history of schizophrenia.

Twentieth-century Schizophrenia

The fact that schizophrenia survived critical assault may yet be seen as one of the great intellectual tragedies of twentieth-century psychiatry. It may well yet prove to be psychiatry's phlogiston, and join neurasthenia, neurosis, hysteria, drapteomania, and whatnot in the graveyard of psychiatric concepts. Nevertheless, its survival in the twentieth century was not entirely incomprehensible. At no time in the history of the concept, even during its most psychogenic years, has there ever been an absence of biological and genetic findings. Such data hinted at possible underlying biological deficits related to the kinds of madness, which refused to disappear with a sociological wand. Frequently, methodological and technological advances also seemed promising. These never translated into reliable and valid diagnostic markers for schizophrenia (clearly, shifting classification did little to help alleviate this basic difficulty). Yet their continued historical emergence, alongside a litany of sometimes promising therapeutic endeavours and advances in brain research and statistics, served to shore up faith in an underlying biological basis to madness. It is hard not to have some sympathy for those who associated schizophrenia with this expanding knowledge base. For many, even if it rarely stood up to exacting scrutiny, schizophrenia clearly gave the plausible appearance of insight. It served as a scaffold of sorts and endorsed a groping pragmatism. It contextualised theoretical arguments. And, at least latterly, the concept's perceived reliability had increased dramatically as the century sought closure. As such, it is easy to understand twentieth-century psychiatry's faith in schizophrenia, and that we are close to a neuropsychological explanation (Gray, 1998). Schizophrenia remains a teleological destination as much as it was a point of investigative departure.

Yet when viewed across decades, twentieth-century schizophrenia had been referenced as a symptom (e.g. Bridges, 1919), a disorder (Sarbin, 1972), a group of disorders (Wing, 1985), a spectrum of disorders (Zubin, 1961), and a syndrome (Gottesman and Shields, 1972) (for Bannister (1971) to specify it as a syndrome or talk of schizophrenias was to articulate a semantic weakness of the concept). Schizophrenia had also been—in the incompatible sense—a disease and a psychogenic disorder. On occasion it had been autism, epilepsy, encephalitis lethargicia, and other conditions. And, at times, schizophrenia persisted, so to speak, 'for lack of anything better, which would be generally acceptable' (Lewis, 1979).

Part of the foundational justification of twentieth-century schizophrenia was the seemingly unscientific proposition that the future, in displaying the subdivisions of schizophrenia, would retrospectively reify the group of diseases called schizophrenia. In this reading, the affirmation of schizophrenia's ontological status (and that of its various symptoms) was one of potentially endless deferral to the future. Schizophrenia announced itself through prophecy. Yet schizophrenia, a word that ultimately veiled many conflicting ideas, never had a clear boundary to begin with.

Understandably, then, this made it peculiarly impenetrable to logical attack. Because its subdivision remained a job for the future, and because the entities in question, and indeed because their precise symptoms, were unknown, the disappearance of one theorised grouping—say 'simple schizophrenia', or, indeed, any entity pertaining to schizophrenia (even when it results in a lack of historical continuity) did not negate nor undermine the existence of schizophrenia. Hence, when Bleuler abandoned his emphasis on autism, and conjectured dereism, there was no turmoil. Similarly, there was no elegy for *Benomenheit*. With constantly changing definition, with constantly changing classification, it was, for the most part, probably impossible to disprove much, if anything.

Twentieth-century schizophrenia was never undermined by its contradictions. If anything, it appears to have been supported by them. A good example occurs concerning its supposed absence. From 1929 to 1937 three notable reports were made about an inability to find schizophrenia. In 1929, Seligman reported that in a so-called Stone Age population of New Guinea he saw no psychosis in the villages among natives leading their own lives. This was in contrast to those in close contact with European settlers. Similarly, in Brazil, Lopez reported no schizophrenia among 'true primitives' of the interior of Brazil. And Farris, who spent years among the Bantu in the Congo, also reported no cases of schizophrenia. Moreover, natives had no comprehension of schizophrenia when its symptoms were described. This caused no loss of sleep for schizophrenia researchers. Instead, by 1973, such an *absence* was proposed as *evidence* in support of a viral theory of schizophrenia (Torrey, 1973). As Noll has observed, the source of the concept's power appears to have been 'its ability to expand and contract the parameters of inexplicable madness, to simultaneously take on all meanings, and no meaning, all at the same time' (2011, p. 269). Arguably then, schizophrenia's lack of homogeneity further allowed it to thrive across the fractured, eclectic, and schismatic theoretical landscape that characterised the behavioural professions of the twentieth century.

At times, then, schizophrenia had appeared as a chameleon that reigned supreme, unimpeded by theoretical difficulties. In fact, in many senses, conflicting theoretical ideas were given an otherwise impossible unity by the use of the term schizophrenia. Accompanied at times by a 'progressive' or 'evolutionary' narrative, it was the master of everyone, the servant of all. For those who demurred in North America, the fact that DSM usage facilitated compensation by insurance companies, Food and Drug Administration approval, and grant funding may have made them act otherwise. Some certainly thought so; in 1979, we find the observation that the very 'existence of schizophrenia grants ... conferences, and manuals provides considerable reification for a major psychiatric conceptualisation that many believe to be illogically derived' (Corning and Steffy, 1979, p. 296). Such complaints echoed old ones. As early as 1934 Lewin had argued that,

> The modern psychiatrist is in a very uncomfortable situation. He knows that the theories of the organic etiology of schizophrenia are

> useless and probably fantasies, and that they have been upheld so long and so tenaciously chiefly because of academic pressure, authority and propaganda, and not because of scientific evidence (1934, p. 322).

However, complicating matters for dissenters was the fact that the possibility of demoting the concept had largely passed from authoritative individuals to powerful communal institutions. Such bodies were less than keen to deconstruct their claims to objectivity, and typically they did not promote heretics to their tables on high.

By 1987, the concept's defenders claimed ever greater reliability for the concept of schizophrenia. In comparison with what had preceded it, this was rightly hailed as a triumph of sorts. Yet our examination of the scientific literature shows that even late twentieth-century schizophrenia continued to be shaped by social forces. The literature notes that diagnostic criteria were modified to make them 'friendly'. It notes that compromises were necessary. It notes that agreement could not be found. Schizophrenia was, furthermore, an abstraction whose historical variance and contradictions—such as those found with catatonia—were frequently set aside. This is not surprising. Collective representations 'often impute to the things to which they refer properties that do not exist in them in any form or to any degree whatsoever' (Durkheim, 1995, p. 229). However much it might have aspired to do otherwise, collaborative twentieth-century psychiatry formulated such collective representations. Schizophrenia was an abstraction where mythology went unrecognised, and which could readily be changed with the nosological scalpel.

On face value, twentieth-century schizophrenia did not fulfil what T.E. Wilkerson called the first condition for membership of a *natural kind*. This was the demonstration of real essence, a property or set of properties necessary and sufficient for membership of the kind (Wilkerson, 1995). As noted in our discussion of definition, the most ingenious method of squaring this circle was to reconfigure schizophrenia retrospectively to be, as Meehl once opined, an open concept, or variants of this idea such as a Roschian concept. Such formulations essentially allow schizophrenia to be seen as intrinsically 'fuzzy' or its various incarnations to be seen as a series of familial resemblances. Similarly, some researchers sought to 'tolerate the Protean qualities of schizophrenia, to eschew the Procrustean, and to embrace the Promethean, whatever their source' (Gottesman and Shields, 1967, p. 204). Or that '"schizophrenia", like "dropsy", may indeed, in course of time, become redundant, but nevertheless describes something real'

(Skrabanek et al., 1977, p. 893). By definition, such attitudes trivialised and downplayed the importance of findings that failed to demonstrate conceptual coherence or found difference. Nevertheless, perhaps our historical understanding of 'schizophrenia' is now complicated by changing twentieth-century epistemological beliefs concerning what *constitutes* and what *constituted* conceptual validity. And perhaps the stability of concepts is less important to empiricists that we might ever have imagined. Yet schizophrenia was actually formulated in a historical period that asserted that the validity of a concept rested on the primacy of necessary and sufficient conditions. And, by those standards, these twentieth-century explanations of schizophrenia did little other than to justify and naturalise vagueness in a concept. Such attitudes refused to question its authenticity. And, as such, they refused to ask if schizophrenia had hindered research into twentieth-century madness more than it had facilitated it. In the twenty-first century the concept of schizophrenia is changing yet again. Among other things, subtypes are being abandoned. Where it all leads to we shall see. But that is another history for another historian.

Breaking it Down

For all its conceptual weaknesses, schizophrenia was not 'one of the diseases given over to therapeutic nihilism' (Anon, 1936, p. 1418). As early as 1936, *The Lancet* could speak of four and 20 treatments of schizophrenia. Effects of 'treatment' included incarceration, sterilisation, broken bones, brain damage, and death. Small wonder then that White suspected that some patients with schizophrenia 'got well' from insulin and other therapies in order to escape the sanatorium or at least escape the repetition of treatment (Grob, 1994). As Klerman noted, 'There is hardly an organ of the body that was not excised in the name of therapy' (1978, p.103). And yet when treatment failed, as it frequently did, many were simply confined to the 'back wards'. Here, nurse Alice Robinson (1960), herself advocating communication and love, recalled lines of naked patients, some shaking their fists, shouting, and spitting, as well as semi-assaults. There remains also an unfolding history of psychiatric cemeteries. Of plots filled with the victims of 'misadventures' in medical history.

Although far from omnipresent, such dehumanisation finds a mirror in the literature. Bleuler (1911/1952), for example, would describe his patients diagnosed with schizophrenia working like machines or robots. As would others (Porot et al., 1968). Similarly, for Morgan the catatonic operated 'much the same way as one of those toy engines that you give a push and which keeps going because of the balance wheel in its mechanism' (1928, p. 581). And, 'When he is stimulated from the outside his obedience is that of a machine or a toy' (Morgan, 1928, p.583). While in 1930, Crookshank could write that 'dementia praecox reveals the schizoid type in a sort of chimpanzoid regression ...' (1930, p. 547).

Kalman, in a 1977 confessional anecdote (relating to decades earlier), would illustrate the danger of such dehumanisation when accompanied with therapeutic nihilism:

> A long time ago, in my first year in a department of Psychiatry, we had a catatonic patient, with a severe heart condition, who—because of his heart condition—could not undergo insulin-shock treatment. This was the best treatment available at the time. I suggested to the Head of Department, that the patient should undergo a heart-operation, which in those days was much more dangerous than now. If the operation was successful, we could try insulin-therapy. If not successful, the man would die. 'What has he got to lose? He is catatonic, lying always in bed, unable to move, to speak. He is a cabbage' (1977, p. 201).

Similarly, French psychiatrist Edouard Toulouse would ask, 'why spend so much money to cultivate in warm green houses and to prolong indefinitely the existence of so many idiots and lunatics?' (as cited in Masson and Azorin, 2007, p. 31).

In 1958, Pierre Renchnick observed that at high doses tranquillisers that might suppress certain symptoms of schizophrenia could produce disorganisation in nervous function in normal individuals; there were no miracle pills (Anon, 1958). As early as 1957 their excessive use as sedatives was also being noted in the Swiss popular press (Anon, 1957c). Hurst (1960) further complained that their use in chronic wards was becoming excessive, and that they had replaced barbiturates and paraldehyde as a means of allaying the anxieties of the staff. And by 1977, Bellak would complain of wards full of zombies (Bellak, 1977). For Rollin, their main virtue from a practical standpoint was that such drugs and their recipients were *easy to handle*. They also interrupted acute episodes of schizophrenia (Rollin, 1979). By contrast, one of his 'percipient and articulate' patients reflected, 'madness is preferable to the numbed cabbage I have become' (Rollin, 1979, p. 1775). Another preferred to be 'a little 'mad' than overdosed by major tranquillisers' (ibid). In this context, the 'praecox feeling', or Gruenberg's 1961 definition of a schizophrenic as 'a person whom, after I have talked with him for 15 minutes I consider to be a schizophrenic' (MacMahon, 1961, p. 334), or Lewis and Piotrowski's comment, 'even a trace of schizophrenia is schizophrenia' (cited in Stephens, 1978, p.26), are more than a little disconcerting. As the century closed, serious side effects such as tardive dyskinesia or death remained possible from treatment. That was a high price for a treatment supported by an idea that often failed to justify itself. Hence, for those who seek to understand madness in the twenty-first century, we must continue to interrogate and interpret schizophrenia. We must remain wary of premature synthesis and always seek to break the problem down into more manageable components. There are things other than words at stake.

Epilogue: Consider Nijinsky

In 1939 Vaslav Nijinsky was visited in a Swiss hospital by press photographers.[1] Years earlier Bleuler had assessed the dancer as 'a confused schizophrenic with mild manic excitement' (Acocella, 1999, p. xli). Similarly, H.G. Baynes (1940) would see Nijinsky as so archetypically schizophrenic that he could see the dancer in the drawings of leaping androgynous figures made by others. Although we may question twentieth-century schizophrenia, there is no doubting Nijinsky's troubled mental history—his own writing and diaries bear witness to it. In a garbled and sometimes incoherent letter to Jean Cocteau, written shortly after his diagnosis, signs of Nijinsky's unwellness still echo dancelike across the page: 'Mogi, cogi, togi, jogi. Migi, gigi, gi gi, rigi, Tchigi, tchigi, tchigi, rigi. Tchigi, rigi, rigi, tchigi. Migi, tigi ...' (Acocella, 1999, p. 274). This is tragic, although maybe even delightful in its rhythms for Cocteau, and somewhat fitting for a dancer, who, more than any, was never truly separated from the dance.

Asked by the photographers to reprise his famous jump Nijinsky obliged. The doctor in charge was apparently so angry he threatened to discharge Nijinsky. But records note the applause gave Nijinsky some pleasure (Acocella, 1999). In one of the photographs, later published in *Paris Match* and *Life*, a suited Nijinsky is making his jump. It is not high, but it is photographed so that he appears to levitate—with arms spread out at waist height. Behind him on the wall, he casts a great shadow that reaches to the floor. For this artificing photographer, Jean Manzon, Nijinsky had simultaneously left his split personality behind and revealed it. Yet, for us, he appears to have jumped right out of a twentieth-century concept. Everything and nothing has changed.

Appendix: Goodbye to Hebephrenia

Despite the complex regional variance that is visible throughout the history of schizophrenia's classification, we tend to also find classifiers continuously returned to archetypal 'foundational' schizophrenia subtypes such as hebephrenia. As such, for much of the first half of the twentieth century hebephrenia was fairly easy to find in asylums and their records, unlike catatonia. Yet although hebephrenia was more visible than catatonia, and although it held sway in much of official classification, the history of twentieth-century conceptualisation of schizophrenia reveals a growing lack of confidence in even this core concept. Possibly, its demise has its roots in comments made by Bleuler in 1924, when he admitted that hebephrenia 'now constitutes the big trough into which are thrown the forms that cannot be classed with the other forms' (1916/1924, p. 426). It may also have had its demise in changing asylum conditions, for by 1945 we can find that Rapaport would argue that hebephrenia, like catatonia, was now becoming increasingly rare. It perhaps only represented cases who had become deteriorated and that 'in general relatives do not bring hopeless cases to our hospital' (Rapaport, 1945, p. 19). [The reference to patients not being bought to the hospital possibly also reflects a growing public disillusionment with treatment, or distaste for lobotomy, which would see its usage begin to decline by 1948, prior to the introduction of chlorpromazine (Gelman, 1999).] Rapaport, as such, found room in his study for 'chronic unclassified schizophrenia', 'coarctated preschizophrenia' (marked anxiety, blocking, withdrawal, sexual preoccupation, feeling of strangeness, incompetence, extreme inhibition of affect), 'overideational preschizophrenia' (obsessions, wealth of fantasy, introspection, self-obsession, and preoccupation with own body), deteriorated paranoid schizophrenia, and so forth. But he abandoned the category hebephrenia (Rapaport, 1945).

Rapaport was not alone in having doubts about hebephrenia. For Karl Schneider, simple, catatonic, and paranoid schizophrenia were still usable terms, but 'Hebephrenia is not in the same rank; it is a term related to the age of a person. We count hebephrenia in with simple schizophrenia' (1959, p. 91). Hebephrenia, which often carried with it 'pathoplastically' the features of the period, such as the 'bobby soxer' (e.g. teenage girl; ardent Frank Sinatra fan), was now significantly downgraded by a researcher, whose influence would soon become immense. In 1963, Lorr, Klett, and McNair also downgraded hebephrenia. They identified ten syndromes in schizophrenia, which they reduced to three: 'excitement vs. retardation', 'schizophrenic disorganisation', and 'paranoid process'. In doing so, they discovered that hebephrenia did not match up with the disorganised type as one might have expected, and seemingly excluded it on these grounds (Lorr et al., 1963). Although hebephrenia would make it into the 1968 DSM-II, the subtype now appeared to have few active supporters among key schizophrenia researchers. A later explanation offered by George E. Valliant was that the hebephrenics had simply 'burned out': 'I have interviewed "burned out" hebephrenics who have been hospitalized for decades, but who never have admitted delusions or hallucinations vivid enough for them to be diagnosed "schizophrenic" on the Present State Examination' (1978b, p. 83). Such comments, alongside Schneider and Rapaport's downgrading of hebephrenia, consequently constituted part of a growing consensus that saw hebephrenia as highly problematic, and which would foreshadow a somewhat hushed replacement of hebephrenia in DSM-III with the aptly named 'disorganised schizophrenia'.[1] Although not before Anthony could suggest 'microhebephrenia' (alongside microcatatonia and microparanoia) for transient prepsychotic trends in children (Curran, 1974). This did not seemingly remedy the problem of dumping large number of patients into one category. In a discussion on fads in choosing diagnostic subtypes, John Romano would later report a hitherto unremarked upon 'tendency to resort to chronic undifferentiated schizophrenic subtype as expedient' (1977, p. 533).

Less well known subtypes—when not simple forgotten—could also 'disappear'. In 1976, Manfred Bleuler reported problems with the so-called 'catastrophic schizophrenia' (stemming from Mauz's 1930's schizocaria or schizokar). It was characterised by very acute onset of a most severe psychosis early in life. It was then followed without any interruption by a severe lifelong chronic psychosis (and as noted by Langfeldt (1937) but found only in individuals with higher education; teachers, theologians and students). By 1977, it, too, had practically disappeared in recent decades (Romano, 1977). That said, by 2010, an editorial in *Acta Psychiatrica Scandinavica* argued that the concept of schizophrenia had failed, and called for its replacement by hebephrenia. Hebephrenia was not a subtype of schizophrenia, read the editorial; it was schizophrenia (Taylor et al., 2010).

Notes

Introduction

1. This work visits the nineteenth century but primarily focuses on the period 1908–87 (i.e. Bleuler to DSM-III-TR). The literature grows exponentially in size and complexity beyond 1990.
2. Process schizophrenia was subdivided into four more types. The atypical states were denoted as 'schizophreniform', 'pseudo schizophrenias', 'symptomatological schizophrenias', 'schizophrenic reaction types', 'or whatever they may be called' (Langfeldt, 1937, p. 189).

1 Schizoidia: The Lexicon

1. Oskar Kohnstamm is strongly associated with schizothymia, as is the term 'sandbank symptom' (Kohnstamm, 1914).

2 The Split Personality

1. Richards (1989, p. 118) argues that the psychological process of physiomorphism lies at the root of such ideas. Borges (1971/11979) also touches on this idea, in his afterword to *The Book of Sand.*
2. Sophie Hedwig Bleuler Wasser née Hedwig Wasser (1869–1940). A trained philologist, she appears to have influenced Eugen Bleuler in various important ways. However, there is no evidence that the term schizophrenia was a marital in-joke.
3. Lewis (1928) earlier coined the term 'castrophrenia' to include the clinical manifestations of thought theft obsessions in schizophrenia, which he linked to the castration complex.

3 Definitions of Schizophrenia

1. Although it would be Lacan who was later praised for his discovery of an ontological flaw in the being of the schizophrenic (Ey et al., 1977).
2. The feeling could be evoked by a passage from Victor Tausk's *Influencing Machine,* according to Sass (1987). Laing (1960/1990) thought it ought to be the audience response to Ophelia when she became psychotic. See also Mauz's and the clinicians' use during diagnosis of the intuitive 'dahinter' (loosely: to suss that which lies behind something) (Langfeldt, 1937, p. 30).
3. While Bowman and Kasanin could add the term 'constitutional schizophrenia' for a psychosis of insidious onset occurring in a family with a definite history of mental disease (Bowman and Kasanin, 1933).

4 Catatonia: Faces in the Fire

1. Stoddart would reference catatonia as a symptom and katatonia as a disease, through spelling variation, but such usage is not consistent across the literature. Stoddart also dubiously writes of a characteristic handshake in dementia praecox.
2. Kraepelin would translate *Verrücktheit* as 'paranoia'. This possibly reflected comments made by Hecker on the difficulty of defining *Verrücktheit*—and the collapsing of paranoia into *Verrücktheit* by Kahlbaum (Hecker, 1871; Kraepelin, 1887, p. 324).
3. Freeman, who mentions Dodson, would accept katatonia.
4. In 1877 we can find Hecker asking visiting colleagues to make predictions on fresh cases. But, where Katatonie was suspected, finding such predictions less accurate than those of himself and Kahlbaum, which were based on insights on disease course (Hecker, 1877, p. 604). Hecker lauds the methodology as key to the derivation of both Hebephrenia and Katatonie. Hence, comparing results against those of unsuspecting colleagues would probably have helped convince both Kahlbaum and Hecker of the validity of both their clinical methodology and Katatonie.
5. Fink and Taylor would synonymise tödliche katatonie as 'Bell's mania, pernicious catatonia, lethal catatonia, malignant catatonia, manic delirium, delirious mania, syndrome malign, acute or fulminating psychosis, fatal catatonia, mortal catatonia, catatonic delirious state, hypertoxic schizophrenia, drug-induced hyperthermic catatonia, confusocatatonia, delirium acutum, delire aigu, and exhaustion syndrome' (2003, p. 40).
6. The history of schizophrenia in India has yet to be told. Schizophrenia treatment with sulfur injections can be found as early as 1931 in the Ranchi Indian Mental Hospital, Patna (Dhunjiboy and Bomb, 1931).

5 Chasing the Phantom: Classification

1. Claude is also associated with the term 'schizonévrose' (Braconnier, 2006), and 'schizophrénie larvée' or larval schizophrenia (Claude, 1937).
2. In 1930 Sullivan, referencing prison rape, would, however, speak of a 'form of schizophrenia called Acute Homosexual Panic' (1962, p. 209).
3. The literature is inconsistent in its use of terms like 'groups', 'forms', 'types', and 'subtypes'. Several of the citations in this book will reveal this inconsistency.
4. Crow (1980) himself hypothesised symptoms as reversible and dopamine related (type I), or not (type II).
5. Beck and colleagues would give their own six schizophrenias (types S1, S2, S3, SG, SR1 and SR2) (Beck, 1954).
6. Acute schizophrenia was elsewhere described as a *contradictio in adjecto* (Porot et al., 1968).
7. The pseudonyms of the four girls were Nora, Iris, Myra, and Hester (as in NIMH). The surname Genain was derived from Greek and, according to Rosenthal, meant 'dire birth' or 'dreadful gene', although no genetic evidence was ever supplied to confirm this supposition.

8. Institutional conditions didn't always facilitate clarity: 'Because of their passivity, nudity, and open masturbation, the schizophrenic children were the victims of frightening sexual and sadistic attacks by aggressive children' (Falstein and Sutton, 1958, p. 667).
9. In 1977, Petr Skrabanek would declare in the *British Medical Journal* that 'symptomatic schizophrenia' was a contradiction in terms; if schizophrenia was a shorthand term for various groups of symptoms, then the concept of schizophrenia as an entity would be redundant (Skrabanek et al., 1977).

6 Myth and Forgetting: Bleuler's 'Four As'

1. Most of the writings of European psychiatrists such as Kraepelin and Bleuler remain misunderstood and untranslated. Not only has Anglophone psychiatry not learned from history, it appears it does not want to.
2. Aubrey Lewis once complained that Ugo Cerletti—of electroconvulsive therapy fame—did not think that a woman in a ward was schizophrenic because she didn't satisfy Kraepelin's test obeying the command to stick out her tongue so that a pin could be stuck in it (Angel et al., 2003, p. 101).

7 Social Prejudice

1. Similarly in 1975, one psychoanalytic interpretation of schizophrenia could state that the schizophrenic is or becomes schizophrenic by virtue of his own desire (Ey et al., 1977).
2. How serious were such ideas taken? Philip May, speaking of schizophrenia in combat or in *basic training* camps, noted that 'Characteristically, in such situations we see "3, 4-, and 5-day schizophrenia" ... perhaps I step out of line a bit, but I am not exaggerating all that much' (Brill et al., 1969, p. 122).
3. Famously, for Foucault, *'Là où il y a œuvre, il n'y a pas folie'* (1961/2006, p. 117).

8 Contesting Schizophrenia?

1. Leonard Cohen would even tour asylums on LSD.
2. An unimpressed Ey would describe the work as an exercise in 'psychiatricide' (Roudinesco, 1986), although it chimes well with the French phenomenological tradition in schizophrenia research. Laing's work would also be described as 'antinosographique' (Ey et al., 1977, p. 196).
3. In 1936, for example, Becker had insisted on the validity of compartmentalisation in sociology. Resisting charges of 'cultural schizophrenia' made in the journal *Social Forces*, Becker urged the creation of bigger and better 'schizophreniacs'. Becker wanted sociologists in particular to be 'sane schizoids' and crazy 'in a particular way' (1936, p. 104).
4. I can't find where he called it a sacred cow, as some have written.
5. n.b. Medicalisation appears to have occurred alongside deinstitutionalisation, rather than caused deinstitutionalisation (Gelman, 1999). The actual decline in numbers began in the mid-1940s, after the Second World War confirmed

the importance of environmental factors in mental health (Grob, 1994; Gunderson et al., 1974).

6. By 1990, Robert Plomin would still admit that there was no way to explain the substantial discordance for identical twins for schizophrenia 'as currently diagnosed' (1990, p. 188), other than by nongenetic factors.
7. Arguably, the best of antipsychiatry would later reflower in patient advocacy movements.

9 Manufacturing Consensus in North America

1. Latent schizophrenia didn't last long. By 1974 latent schizophrenia was ominously declared (although prematurely) to have been abandoned (Calanca, 1974). And in a 1976 critique of one study, Theodore Lidz declared, 'I would not consider it a methodologically secure measure to include borderline and latent schizophrenia', as both categories (and in addition 'pseudomutual schizophrenia') were 'highly arbitrary diagnoses' (1976, p. 406).
2. Its section on disorders simultaneously constituted a subset of the American Medical Association's (1937) fourth edition of the *Standard Nomenclature of Diseases and Operations*.
3. These were listed as disturbances of association, thought disorder, changes in affectivity, a tendency to prefer fantasy to reality, and to seclude oneself from reality, and autism.
4. These critiques were generally powerful, although lacking much historical emphasis.

Epilogue

1. The visit occurred after a series of insulin shock treatments were said to have alleviated his illness (in fact, they apparently exacerbated it).

Appendix

1. In DSM-III, absence of systematised delusions, which most of us would hope to show, was one of three important criteria in the checklist for disorganised type. The choice of name might have been influenced article by Seymour Epstein (1979) in the *Schizophrenia Bulletin*, who discussed the possibility that schizophrenic disorganisational states are the consequence of a natural adaptive process, albeit a desperate one.

Further Reading

Adeoye Lambo, T. (1960) Neuropsychiatric problems in Nigeria. *British Medical Journal*, 2, 1696–704.

Adityanjee, Aderibigbe, Y.A., Theodoridis, D. and Vieweg, V.R. (1999) Dementia praecox to schizophrenia: The first 100 years. *Psychiatry and Clinical Neurosciences*, 53, 437–48.

American Medico-Psychological Association and National Committee for Mental Hygiene (1920) *Statistical Manual for the Use of Institutions for Mental Diseases* (Utica, NY: State Hospitals Press).

American Medico-Psychological Association and National Committee for Mental Hygiene (1927) *Statistical Manual for the Use of Hospitals for Mental Diseases* (New York: Boyd Printing Company).

Andreason, N.C. (1982) Negative symptoms in schizophrenia: Definition and reliability. *Archives of General Psychiatry*, 39, 784–8.

Anon. (1923) *Travaux de L'Université de Strasbourg* (Strasbourg: Alsacienne).

Anon. (1925) *Travaux de L'Université de Strasbourg* (Strasbourg: Alsacienne).

Anon. (1931) Explaining the weird Kansas Scarecrow. *Ogden Standard Examiner*, Dec 13, p. 31

Anon. (1935a) London association of the medical women's federation. Some common mental disorders. *The Lancet*, 225, 327.

Anon. (1935b) Strictly speaking. *Washington Post*, Jan 15, p. 12.

Anon. (1941a) Hess viewed as 'Jekyll-Hyde'. *New York Times*, May 19, p. 4.

Anon. (1941b) Shock absorber. *The Washington Post*, Sep 12, p. 10.

Anon. (1944) Hopitaux psychiatriques. *Revue des Établissements et des Oeuvres de Bienfaisance*, 1, 39–41.

Anon. (1951) Review of schizophrenic art. *Journal of Consulting Psychology*, 15, 87.

Anon. (1952) Attack schizophrenia. *The Science News Letter*, 61, 99.

Anon. (1954) Review of the six schizophrenias. *Journal of Consulting Psychology*, 18, 472–3.

Anon. (1955) Any questions? *British Medical Journal*, 1, 117–18.

Anon (1957). Story of schizophrenia, Lizzy plays at Wayside. *Winnipeg Free Press*, Feb 15, p. 11

Anon. (1959a) Au Pakistan, le chimiste se penche sur son passé. *Journal de Genève*, Jun 24, p. 6.

Anon. (1959b) Mothers of schizophrenics. *British Medical Journal*, 1, 102–3.

Anon. (1959c) The new pictures. *Time Magazine*, Mar 23. Available at: http://www.time.com/time/magazine/article/0,9171,892387,00.html (accessed 15 May 2007).

Anon. (1965) Clin-Alert: March 23. *Clin-Alert*, 3, 30–4.

Anon. (1968) 'Schizophrenia', in: *Encyclopedia Britannica*, Vol. 19, p. 1162 (London: William Benton).

Anon. (1974) At issue: a balance of diagnostic deficit. *Schizophrenia Bulletin*, 1, 4.

Anon. (1977) Torque test for schizophrenia. *Science News*, 112, 167.

Anon. (1995) We Ssay Syrah, they say Shiraz. *Washington Post*, Jan 25, p. 9.

Anon. (2005) 'Schizophrenia', in: *Oxford English Dictionary*. Available at: http://www.oed.com (accessed 25 August 2005).

Arieti, S. (1979) *Understanding and Helping the Schizophrenic. A Guide for Family and Friends* (New York: Touchstone Books).

Arieti, S. (1981) *Understanding and Helping the Schizophrenic. A Guide for Family and Friends* (London: Pelican).

Ballet, G. (1894) 'Psychoses', in: J.M. Charcot, C.J. Bouchard and E. Brissaud (eds) *Traité de médicine, tome VI*, pp. 1057–206 (Paris: Masson).

Barbier, J.M., Serra, G., Loas, G. and Breathnach, C.S. (1999) Constance Pascal: Pioneer of French psychiatry. *History of Psychiatry*, 10, 425–37.

Barrett, A. (1924) Constitution and disposition in psychiatric relations. *American Journal of Psychiatry*, 81, 246–60.

Barrett, A.M (1925/1928) 'Heredity relations in schizophrenia', in: C.L. Dana, T.K. Davis, S.E. Jelliffe, H.A. Riley, F. Tilney and W. Timme (eds) *Schizophrenia [Dementia Praecox]; An Investigation by the Association for Research in Nervous and Mental Diseases*, pp. 67–99 (New York: Paul B Hoeber Inc.).

Bates, C. and Van Dam, C. (1979) Low incidence of schizophrenia in British Columbia coastal Indians. *Journal of Epidemiology and Community Health*, 38, 127–30.

Bellows, M. (1915) *Dictionary of German and English* (London: Longmans, Green & Company).

Benediktson, D.T. (1989) Caligula's madness: madness or interictal temporal lobe epilepsy? *The Classical World*, 82, 370–5.

Berrios, G. (1999) Classifications in psychiatry: A conceptual history. *Australian and New Zealand Journal of Psychiatry*, 33, 145–60.

Bertillon, J. (1910/2008) *International Classification of Causes of Sickness and Death* (Whitefish, MT: Kessinger Publishing's Rare Reprints).

Birley, J.L. (2002) Famine: the distant shadow over French Psychiatry. *British Journal of Psychiatry*, 180, 298–9.

Black, D.W. and T. Boffeli (1989) Simple schizophrenia: Past, present, and future. *American Journal of Psychiatry*, 146, 1267–73.

Bleuler, E. (1906/1969) 'Significance of association experiments (M.D. Eder, trans.)', in: C.G. Jung (ed.) *Studies in Word-association*, pp. 1–7 (London: Routledge and Keegan Paul).

Bleuler, E. (1910) Die psychoanalyse Freuds [Freudian psychoanalysis]. *Jahrbuch für Psychoanalytische und Psychopathologische Forschungen*, 2, 40–1.

Bleuler, E. (1921) Kretschmer: Korperbau und character. *Munchener Medizinische Wochenschrift*, 33, 1057–9.

Bleuler, E. (1922) Die Probleme der Schizoidie und der Syntonie [The problems of the schizoid and the syntonic]. *Zeitschrift für die Gesamte Neurologie und Psychiatrie*, 78, 373–99.

Bleuler, M. (1979) On schizophrenic psychoses. *American Journal of Psychiatry*, 136, 1403–9.

Blumenthal, A.L. (1975/2002) 'A reappraisal of Wilhelm Wundt', in: W. Pickren and D. Dewsbury (eds) *Evolving Perspectives on the History of Psychology*, pp. 65–78 (Washington, DC: American Psychological Association).

Boring, E.G. (1913a) Introspection in dementia praecox. *American Journal of Psychology*, 24, 145–70.

Boring, E.G. (1913b) The course and character of learning in dementia praecox. *Government Hospital for the Insane Bulletin*, 5, 51–80.

Boring, E.G. (1929) *A History of Experimental Psychology* (London: Appleton Century Co.).

Bottas, A., Cooke, R.G. and Richter, M.A. (2005) Comorbidity and pathophysiology of obsessive-compulsive disorder in schizophrenia: Is there evidence for a schizo obsessive subtype of schizophrenia? *Journal of Psychiatry and Neuroscience*, 30, 187–93.

Breisach, E. (1994) *Historiography. Ancient, Medieval, and Modern* (Chicago, IL: University of Chicago Press).

Brissaud, E., Pinard, A. Reclus, P. Couvelaire, A., Lenormant, C. and Meige, H. (1912) *Nouvelle Pratique Médico-Chirurgicale* (Paris: Masson).

Brock, A. (1993) Something old, something new: The 'reappraisal' of Wilhelm Wundt in textbooks. *Theory and Psychology*, 3, 235–42.

Brock, A. (2006) Rediscovering the history of psychology: Interview with Kurt Danziger. *History of Psychology*, 9, 1–16.

Bührmann, M.V. (1982). Thwasa and bewitchment. *South African Medical Journal*, 61, 877–9.

Buss, A.H. (1978) *Psychology, Behaviour in Perspective* (Hoboken, NJ: John Wiley).

Bynum, W.F. (1994) *Science and the Practice of Medicine in the Nineteenth Century* (Cambridge: Cambridge University Press).

Calkins, M.W. (1901) *An Introduction to Psychology* (London: Macmillan).

Campbell, C. (1926) Review of physique and character: An investigation of the nature of constitution and of the theory of temperament. *Journal of Abnormal and Social Psychology*, 21, 107–8.

Canguilhem, G. (1989) *The Normal and the Pathological* (C.R. Fawcett, trans.) (New York: Zone Books).

Carotenuto, A. (1984) *A Secret Symmetry: Sabina Spielrein Between Jung and Freud* (London: Routledge & Kegan Paul).

Carpenter, W.T. and Stephens, J.H. (1979) An attempted integration of information relevant to schizophrenic subtypes. *Schizophrenia Bulletin*, 5, 490–506.

Casier, R. (2005) Self portrait. *Schizophrenia Bulletin*, 31, 183.

Charcot, J.M., Bouchard C.J. and Brissaud, E. (1894) *Traité de médicine, tome VI* (Paris: Masson).

Chartier, R. (1989/1995) 'The world as representation', in: J. Revel and L. Hunt (eds) *Histories: French Constructions of the Past*, pp. 543–88 (New York: The New Press).

Chaslin, P. (1895) *La confusion mentale primitive, stupidité, démence aiguë, stupeur primitive* (Paris: Asselin et Houzeau).

Ciompi, L. (1988) *The Psyche and Schizophrenia. The Bond Between Affect and Logic* (D.L. Schneider, trans.) (Cambridge, MA: Harvard University Press).

Claude, H. (1930) Schizomanie a forme imaginative [The imaginative form of schizomania]. *Encephale*, 25, 715–27.

Claude, H., Borel, A. and Robin, G. (1924) Une automutilation révelatrice d'un état schizomaniaque [A revealing automutilation in a schizomanic state]. *Annales Medico-Psychologiques*, 1, 331–9.

Cocks, G. (1994) 'German Psychiatry during the nazi period', in: M. Micale and R. Porter (eds) *Discovering the History of Psychiatry*, pp. 282–307 (Oxford: Oxford University Press).

Cooksy, E.C. and Brown, P. (1988) Spinning on its axes: DSM and the social construction of psychiatric diagnosis. *International Journal of Health Services*, 28, 525–54.

Cooper, B. (1978) 'Epidemiology', in: J.K. Wing (ed.) *Schizophrenia: Towards a New Synthesis*, pp. 31–53 (London: Academic Press).

Coritat, I. (1909) Certain pulse reactions as a measure of the emotions. *The Journal of Abnormal Psychology*, 4, 261–79.

Coriat, I. (1913/1916) *Abnormal Psychology* (New York: Moffat, Yard & Company).

Cornell, E.L. and Lowden, G.L. (1923) A comparison of the Stanford and Porteus tests in several types of social inadequacy. *The Journal of Abnormal Psychology and Social Psychology*, 18, 33–42.

Coulter, J. (1972) *Approaches to Insanity. A Philosophical and Sociological Study* (Bath: The Pitman Press).

Cronbach, L.J. and Meehl, P.E. (1955) Construct validity in psychological tests. *Psychological Bulletin*, 52, 281–302.

Crossley, N. (2006) *Contesting Psychiatry: Social Movements in Mental Health* (New York: Routledge).

Crow, T.J. (1993) Two syndromes in schizophrenia and their underlying pathology. *Clinical Medicine*, 48, 9.

Dabout, E. and Roussy, G. (1924) *Petit dictionnaire de médecine* (Paris: Baillière et Fils).

Daco, P. (1971) *Guide de la psychologie moderne* [Guide to modern psychology] (Paris: Éditions Robert Langeac).

Danziger, K. (1990) *Constructing the Subject* (New York: Cambridge University Press).

Decker, H.S. (2007) How Kraepelinian was Kraepelin? How Kraepelinian are the neo-Kraepelinians? From Emil Kraepelin to DSM-III. *History of Psychiatry*, 18, 337–60.

De Fleury (1927) Nomination de deux délégués de l'academie dans le comité d'une foundation anonyme. *Bulletin de l'Académie nationale de médecine*, 98, 32–3.

De Jong, H. (1946) Experimental catatonia in rats produced by centrifugation. *The Journal of Comparative Psychology*, 38, 17–26.

Deleuze, G. (2006) *Two Regimes of Madness. Texts and Interviews 1975–1995* (New York: Semiotexte).

Devereux, G. (1939) A sociological theory of schizophrenia. *Psychoanaltic Review*, 26, 315–42.

Drever, J. (1952) *The Penguin Dictionary of Psychology* (Harmondsworth: Penguin).

Drever, J. (1952/1964) *The Penguin Dictionary of Psychology*, rev. ed. (Harmondsworth: Penguin).

Ehrenwald, J. (1948) *Telepathy and Medical Psychology* (New York: W.W. Norton).

Eidelberg, L. (ed.) (1968) *Encyclopedia of Psychoanalysis* (London: The Free Press).

Eliot, T.S. (1933) *The Use of Poetry and the Use of Criticism* (London: Faber).

Emery, H.G. and Brewster, K.G. (1927) *The New Century Dictionary of the English Language* (London: The Century Co.).

Eure-et-Loir. Conseil général (1938) *Département d'Eure-et-Loir, Conseil général* (Chartres: Lainé et Tantet). Available at: http://gallica.bnf.fr/ark:/12148/bpt6k57116982/f1.image.r=bouffee%20schizophrenie.langEN (accessed 27 February 2015).

Faris, R.E.L. (1934) Some observations on the incidence of schizophrenia in primitive societies. *The Journal of Abnormal and Social Psychology*, 29, 30–1.

Faris, R.E.L. and Dunham, H.W. (1967) *Mental Disorders in Urban Areas. An Ecological Study of Schizophrenia and Other Psychoses* (Chicago, IL:University of Chicago Press).

Finke, M. (1999) Review of the book Freud and the Bolsheviks: Psychoanalysis in imperial Russia and the Soviet Union. *The Slavic and East European Journal*, 43, 750–1.

Fish, F. (1957) The classification of the schizophrenia. *Journal of Mental Science*, 103, 443–65.

Fish, F. (1967/1977) 'Clinical presentation and classification of schizophrenia', in: T. Silverstone and B. Barraclough (eds) *Contemporary Psychiatry. Selected Reviews From the British Journal of Hospital Medicine*, p. 3 (Kent: Headly Brothers).

Foucault, M. (1972) *The Archaeology of Knowledge* (A.M. Sheridan Smith, trans.) (London: Tavistock Publications).

Fowles, D.C. (1984) 'Biological variables in psychopathology. A psychobiological perspective', in: H. Adams and P. Sutker (eds) *Comprehensive Handbook of Psychopathology*, pp. 77–111 (London: Plenum Press).

François, A. (1919) Analyse psychiatrique des confessions des Jean-Jacques Rousseau par V.Demole. *Annales De La Société J.J. Rousseau*, 12, 184–5.

Frazier, S.H. and Carr, A. (1964) *Introduction to Psychopathology* (London: Collie Macmillan).

Freud, S. (1911/1981) 'Psychoanalytic notes on an autobiographical account of a case of paranoia (dementia paranoides)', in: J. Strachey and A. Freud (eds) *The Standard Edition of the Complete Psychological Works of Sigmund Freud*, vol. xii, pp. 75–6 (London: Hogarth Press).

Freud, S. (1914) *The History of the Psychoanalytic Movement* (New York: Nervous and Mental Disease Publishing).

Freud, S. (1936) Métapsychologie. *Revue Française de Psychanalyse*, 9, 22–116.

Garrison, V. (1978) Support systems of schizophrenic and nonschizophrenic Puerto Rican migrant women in New York city. *Schizophrenia Bulletin*, 4, 561–96.

Gilyarovsky, V.A. (1985) 'The Soviet Union', in L. Bellak (ed.) *Contemporary European Psychiatry*, pp. 283–327 (New York: Grove Press).

Goldenson, R.M. (1984) *Longman's Dictionary of Psychology and Psychiatry* (London: Longman).

Grossmann, E. (1957) Herbs in folk medicine. *Western folklore*, 16, 290–2.

Grube, M. (2006) Towards an empirically based validation of intuitive diagnostic: Rümke's 'praecox feeling' across the schizophrenia spectrum: Preliminary results. *Psychopathology*, 395, 209–17.

Guiraud, P. (1968) 'Origine et évolution de la notion de schizophrénie', in: *Confrontations psychiatriques: Évolution de la schizophrénie*, pp. 9–31 (Paris: Specia).

Guze, S.B. (1980) 'Schizoaffective disorders', in: H. Kaplan, A. Freedman and B. Sadock (eds) *Comprehensive Textbook of Psychiatry*, 3rd ed., Vol. 3, pp. 1301–5 (London: Williams & Wilkins).

Guze, S.B., Olin, S.T., Strauss, J.S., Katz, M.M. and Kendell, R.E. (1982) Comments on Blashfield's article. *Schizophrenia Bulletin*, 8, 6–12.

Hacking, I. (1975) *The Emergence of Probability: A Philosophical Study of Early Ideas About Probability, Induction and Statistical Inference* (London: Cambridge University Press).
Hacking, I. (1983/2005) *Representing and Intervening. Introductory Topics in the Philosophy of Natural Science* (London: Cambridge University Press).
Hacking, I. (1996) Automatisme ambulatoire: Fugue, hysteria, and gender at the turn of the century. *Modernism/Modernity*, 3, 31–43.
Hafner, H. (2004) Schizophrenia: Still Kraepelin's dementia praecox? *Epidemiologia e Psichiatria Sociale*, 13, 99–112.
Hamilton, M. (1967) *Abnormal Psychology* (Harmondsworth: Penguin).
Hammond, W. (1883) *A Treatise on Insanity in its Medical Relations* (New York: Appleton).
Hanfmann, E. (1939) Analysis of the thinking disorder in a case of schizophrenia. *Archives of Neurology and Psychiatry*, 41, 568–79.
Hanfmann, E. and Kasanin, J. (1942) *Conceptual Thinking in Schizophrenia* (New York: Nervous and Mental Disease Monographs).
Haouzir, S., and Bernoussi, A. (2005) *Les schizophrénies* [*The Schizophrenias*] (Paris: Armand Colin).
Hardin, G. (1956) Meaninglessness of the word protoplasm. *Scientific Monthly*, 88, 12–120.
Hardin, G. (1993) *Living Within Limits. Ecology, Economics and Population Taboos* (Oxford: Oxford University Press).
Harding, E.M. (1941) Review of mythology of the soul: A research into the unconscious from schizophrenic dreams and drawings. *The Journal of Abnormal and Social Psychology*, 36, 129.
Hare, E.H. (1998) *On the History of Lunacy: The 19th Century and After* (London: Gabbay).
Harmatz, M.G. (1978) *Abnormal Psychology* (London: Prentice Hall).
Harré, R. (1986) *Varieties of Realism. A Rational for the Natural Sciences* (Oxford: Basil Blackwell).
Harré, R. and Lamb, R. (1983/1986) *The Dictionary of Physiological and Clinical Psychology* (Cambridge, MA: MIT Press).
Harriman, P.L. (ed.) (1946) *Encyclopaedia of Psychology* (New York: The Philosophical Library).
Harris, B. (1974/2002) 'Whatever happened to little Albert?', in: W. Pickren and D. Dewsbury (eds) *Evolving Perspectives on the History of Psychology*, pp. 237–54 (Washington, DC: American Psychological Association).
Hayasaka, C. (1936) Société Psychanalytique de Sendai. *Revue Française de Psychanalyse*, 9, 717–19.
Hayward, A.L. (1949) *Cassell's New English Dictionary*, 15th ed. (London: Cassell and Company).
Hempel, C.G. (1961) 'Introduction to problems in taxonomy', in: J. Zubin (ed.) *Field Studies in the Mental Disorders*, pp. 3–50 (New York: Grune & Stratton).
Hill, G.H. (1900). Dementia praecox. *American Journal of Insanity*, 57, 319–24.
Hitch, M. (1956) *Baillière's Nurses' Medical Dictionary* (London: Baillière Tindall and Cox).
Hoch, P.B. (1961) 'From the point of view of psychiatry', in: J. Zubin. (ed.) *Field Studies in the Mental Disorders*, pp. 342–53 (New York: Grune & Stratton).

Hoffer, A. and Osmond, H. (1974) *How to Live with Schizophrenia* (New York: University Books).

Holzinger, A., Angermeyer, M.C. and Matschinger, H. (1998) What do you associate with the word schizophrenia? A study of the social representation of the word schizophrenia. *Psychiatrische Praxis*, 25, 9–13.

Hornstein, G. A. (1992/2002) 'The return of the repressed: Psychology's problematic relations with psychoanalysis: 1909–1960', in: W. Pickren and D. Dewsbury (eds) *Evolving Perspectives on the History of Psychology*, pp. 471–91 (Washington, DC: American Psychological Association).

Huber, G. (1957) Die coenästhetische Schizophrenie [Cenestho-pathic Schizophrenia]. *Fortschritte der Neurologie-Psychiatrie*, 25, 491–520.

Ikeda, K., Ball, H.V. and Yamamura, D.S. (1962) Ethnocultural factors in schizophrenia in the Japanese in Hawaii. *American Journal of Sociology*, 68, 242–8.

Institut National de la Statistique et des Études Économiques (1964) Annuaire Statistique de la France, 12, 84.

Jackson, D. (1960) *The Etiology of Schizophrenia* (New York: Basic books).

Janet, P. (1921) The fear of action. *The Journal of Abnormal Psychology and Social Psychology*, 16, 150–60.

Jenner, F.A. (1971) Dissenters in Soviet mental hospitals [Letter to the Editor]. *The Times*, Sep 15, p. 17.

Johnson, L.C. (1949) Wechsler-Bellevue pattern analysis in schizophrenia. *Journal of Consulting Psychology*, 13, 32–3.

Jones, H.G. (1978) 'Psychological aspects of treatment of in-patients', in: J.K. Wing (ed.) *Schizophrenia: Towards a New Synthesis*, pp. 189–211 (London: Academic Press).

Jung, C. (1914/1972) 'The content of the psychosis', in: W. McGuire (ed.) *The Collected Works of C.G. Jung*, Vol. 3, pp. 153–79 (London: Routledge & Keegan Paul).

Jung, C. (1969) *Studies in Word-association* (M.D. Elder, trans.) (London: Routledge & Keegan Paul).

Jung, C. (1974) 'Letter to Sigmund Freud. May 13th, 1907', in: W. McGuire (ed.) *The Freud/Jung Letters: The Correspondence Between Sigmund Freud and C.G. Jung*, pp. 43–5 (Princeton, NJ: Princeton University Press).

Jung. C. (1974b) 'Letter to Sigmund Freud. April 30th 1910', in: W. McGuire (ed.) *The Freud/Jung Letters: The Correspondence Between Sigmund Freud and C.G. Jung*, p. 312 (Princeton, NJ: Princeton University Press).

Kempf, E.J. (1948) Bisexual factors in curable schizophrenia. *The Journal of Abnormal and Social Psychology*, 44, 414–19.

Kesey, K. (1962) *One Flew Over the Cuckoo's Nest* (London: Methuen).

Klein, M. (1988) *Envy and Gratitude and Other Works 1946–1963* (London: Vintage).

Kleinman, J., Gillin, J. and Wyatt, R. (1977) A comparison of the phenomenology of hallucinogens and schizophrenia from some autobiographical accounts. *Schizophrenia Bulletin*, 3, 560–86.

Kleist, K. (1923) Die Auffassung der Schizophrenie als psychische Systemerkrankungen (Heredodegenerationen) [The view of schizophrenia as a system of mental illness]. *Klinische Wochenschrift*, 21, 962–3.

Knight, R.P. (1951) 'Introduction', in: E. Brody and F. Redlich (eds) *Psychotherapy with Schizophrenics*, p. 11 (New York: International Universities Press).

Kraepelin, E. (1899) *Psychiatrie: Ein lehrbuch für studirende und aerzte* [A Textbook for Students and Doctors], 6th ed. (Leipzig: Barth).

Kuhn, T. (1996) *The Structure of Scientific Revolutions*, 3rd ed. (Chicago, IL: University of Chicago Press).

Kuiper, A.M. (2007) Losing reality. *Schizophrenia Bulletin*, 33.

Kula, A. (1998) *Studies in Scientific Realism* (Oxford: Oxford Universities Press).

Laboucarié, J. (1968) 'Le devenir des psychoses délirantes aiguës et le risque de leur evolution schizophrénique secondaire', in: *Confrontations psychiatriques: Évolution de la schizophrénie*, pp. 31–52 (Paris: Specia).

Lacan, J. (1932) *De la psychose paranoïaque dans ses rapports avec la personnalité* (Paris: Éditions du Seuil).

Laforgue, R. (1926) Schizophrenia, schizomania and schizonoïa. *Zeitschrift für die Gesamte Neurologie Und Psychiatrie*, 105, 448–58.

Laforgue, R. (1927) Scotomization in schizophrenia. *International Journal of Psycho-analysis*, 8, 473–9.

Laird, D.A. (1928) Reading character may soon be possible. *The science News-letter*, 14, 44–50.

Lamont, A. and Moffson, A. (1954) Forensic psychiatry at Weskoppies mental hospital. *South African Medical Journal*, 28, 372–6.

LeGoff, J. (1974/1995) 'Mentalities: A history of ambiguities', in: J. Revel and L. Hunt (eds) *Histories: French Constructions of the Past*, pp. 241–7 (New York: The New Press).

Leighton, A.H. (1961) 'From the point of view of social psychiatry', in: J. Zubin (ed.) *Field studies in the Mental Disorders*, pp. 324–32 (New York: Grune & Stratton).

Le Maître, P. (1895) *Contribution à l'étude des états cataleptiques dans les maladies mentales* (Paris: G. Steinheil).

Levin, L.S. (1949) The utility of Wechsler's patterns in the diagnosis of schizophrenia. *Journal of Consulting Psychology*, 13, 28–31.

Lewis, N.D.C. (1928) The psychobiology of the castration reaction. *The Psychoanalytic Review*, 15, 53–85.

Lewis, N.D.C. and Piotrowski, Z.A. (1955) 'Clinical diagnosis of manic-depressive psychosis', in: P. Hoch and J. Zubin (eds) *Depression*, pp. 25–38 (New York: Grune & Stratton).

Littlewood, R. (1986) Anthropology and British psychiatry. *Anthropology Today*, 2, 8–11.

Lomas, J. (1957) Treatment of schizophrenia. Pactal and chloropromazine compared. *British Medical Journal*, 2, 78–80.

Lundholm, H. (1922) Reaction time as an indicator of emotional disturbances in manic depressive psychoses. *The Journal of Abnormal Psychology and Social Psychology*, 17, 292–318.

MacDonald, A.M. (ed.) (1972) *Chamber's Twentieth Century Dictionary* (London: W & R. Chambers).

MacDonald, R. and Shields, W. (1959) Trifluoperazine dihydrochloride ('dtelazine') in paranoid schizophrenia. *British Medical Journal*, 1, 549–50.

Macey, D. (1993) *The Lives of Michel Foucault* (London: Hutchinson).

McGlashan, T.H. and Fenton, W.S. (1994) The positive–negative distinction in schizophrenia. Review of natural history validators. *Arch Gen Psychiatry*, 49, 63–72.

McNeil, E.B. (1971) *The Psychoses* (Upper Saddle River, NJ: Prentice Hall).

McNiff, S. (1974) The myth of schizophrenic art. *Schizophrenia Bulletin*, 1, 12–13.

Magaret, A. and Wright, C. (1943) Limitations in the use of intelligence test performance to detect mental disturbance. *Journal of Applied Psychology*, 27, 387–98.

Magnan, V. (1893) *Recherches sur les centres nerveux. Pathologie et physiologi-pathologique* (Paris: Masson).

Malmo, R.B. (1952) Review of schizophrenic art: its meaning in psychotherapy. *Canadian Journal of Psychology*, 6, 46–7.

Mannheim, K. (1925/1986) *Conservatism. A Contribution to the Sociology of Knowledge* (D. Kettler and V. Meja, trans.) (London: Routledge & Kegan Paul).

Medvei, V. (1993) *The History of Clinical Endocrinology: A Comprehensive Account of Endocrinology from Earliest Times to the Present Day* (Lancashire: Taylor & Francis).

Meyer, A. (1917) The aims and means of psychiatric diagnosis. *American Journal of Psychiatry*, 74, 163–8.

Meyer, A. (1922) Constructive formulation of schizophrenia. *American Journal of Psychiatry*, 1, 355–64.

Meyer, A. (1927) Kraepelin. *American Journal of Psychiatry*, 83, 748–55.

Meyer, J. (1961) 'Problems of nosology and nomenclature', in: J. Zubin (ed.) *Field Studies in the Mental Disorders*, pp. 64–72 (New York: Grune & Stratton).

Midelfort, H.E. (1981) Madness and the problems of psychological history in the sixteenth century. *The Sixteenth Century Journal*, 12, 5–12.

Mignard, M. (1909) Nouvelles. *Revue de Psychiatrie et de Psychologie Expérimentale*, 8, 201.

Mignard, M. (1928) *L'Unité psychique et les troubles mentaux* [Psychic unity and mental disorders] (Paris: Librairie Félix Alcan).

Millon, T. (1986) 'The past and future of DSM-III', in: T. Millon and G.L. Klerman (eds) *Contemporary Directions in Psychopathology. Towards the DSM-IV*, pp. 73–99 (New York: The Guilford Press).

Minkowska, F. (1925) Troubles Essentiels de la schizophrénie. *Evolution psychiatrique*, 127–141.

Montuschi, E. (2003) *The Objects of Social Science* (London: Continuum).

Moore, J.N.P. and Martin, E.A. (1957) Trial of reserpine in treatment of schizophrenia. *British Medical Journal*, 1, 8–14.

Moore, T.V. (1929/1930) The empirical determination of certain syndromes underlying praecox and manic depressive psychoses. *American Journal of Psychiatry*, 86, 719–38.

Morten, H. (1953) *The Nurse's Dictionary*, rev. ed. (London: Faber & Faber).

Murphy, G. (1929) *An Outline of Abnormal Psychology* (New York: The Modern Library).

Neumärker, K. and Bartsch, A.J. (2003) Karl Kleist (1879–1960). A pioneer of neuropsychiatry. *History of Psychiatry*, 14, 441–58.

Nijinsky, R. (1972). *Nijinsky* (New York: Pocket Books).

Nodine, J.H. and Moyer, J.H. (1962) *Psychosomatic Medicine: The First Hahnemann Symposium* (Philadelphia, PA: Lea & Febiger).

Nogué, R. (1927) La schizophrénie. *L'Association Médicale*, 7, 22.

O'Donnell, J.M. (1978/2002) 'The crisis of experimentalism in the 1920s: E.G. Boring and his uses of history', in: W. Pickren and D. Dewsbury (eds) *Evolving Perspectives on the History of Psychology*, pp. 45–56 (Washington, DC: American Psychological Association).

Otsuka, K. and Sakai, A. (2004) Haizmann's madness: The concept of bizarreness and the diagnosis of schizophrenia. *History of Psychiatry*, 15, 73–82.
Pap, A. (1958) *Semantics and Necessary Truth* (New Haven, CT: Yale University Press).
Parnell, R.W. and Skottowe, I. (1959) Length of stay in mental hospitals and some factors influencing it. *British Medical Journal*, 2, 1296–300.
Pascal, C. (1911) *La Démence Précoce* (Paris: Librairie Félix Alcan).
Passmore, R. and Robinson, J.S. (1974) *A Companion to Medical Studies* (London: Blackwell).
Petite, G. (1913). *Essai sur une variété de pseudo-hallucinations, les autoreprésentations aperceptives* (Bordeaux: Cadoret).
Poland J. (2007) 'How to move beyond the concept of schizophrenia', in: M. Chung, K.W. M. Fulford and G. Graham (eds) *Reconceiving Schizophrenia*, pp. 45–56 (Oxford: Oxford: University Press).
Putnam, F.W. (1986) 'The scientific investigation of MPD', in: J.M. Quen (ed.) *Split Minds/Split Brains. Historical and Current Perspectives* (New York: New York University Press).
Radford, J. and Govier, E. (1980) *A Textbook of Psychology* (London: Sheldon Press).
Reddaway, P. (1971) Plea to west on Soviet 'mad-house' jails. *The Times*, Mar 12, p. 8.
Redlich, F.C. and Freedman, D.X. (1966) *The Theory and Practice of Psychiatry* (New York: Basic Books).
Rees, L. (1982) *A Short Textbook of Psychiatry* (Suffolk: The Chaucer Press).
Reid, D.D. (1961) 'Methods and techniques for obtaining comparability in field studies of the mental disorders', in: J. Zubin (ed.) *Field Studies in the Mental Disorders*, pp. 271–324 (New York: Grune & Stratton).
Ritzler, B. and Smith, M. (1976) The problem of diagnostic criteria in the study of the paranoid subclassification of schizophrenia. *Schizophrenia Bulletin*, 2, 209–17.
Roback, A.A. (1962) *History of Psychology and Psychiatry* (London: Vision Press).
Rockmore, T. and Margolis, J. (2006) *History, Historicity and Science* (Aldershot: Ashgate).
Rodin, E.A., Dejong, R.N., Waggoner, R.W. and Bagchi, B.K. (1957) Relationship between certain forms of psychomotor epilepsy and schizophrenia. *Archives of Neurology and Psychiatry*, 77, 449–63.
Róheim, G. (1955). *Magic and Schizophrenia* (London: Indiana University Press).
Rorty, R. (1980) *Philosophy and the Mirror of Nature* (Oxford: Basil Blackwell).
Rosenhan, D.L. and Seligman, M. (1989) *Abnormal Psychology*, 2nd ed. (London: W.W. Norton).
Roubinovitch, J. (1896) *Des variétés cliniques de la folie en France et en Allemagne* (Paris: O.Doin).
Ruckmick, C.A. (1928) *German–English Dictionary of Psychological Terms* (Iowa City, IA: Athens Press).
Rümke, H.C. (1961) 'Problems of nosology and nomenclature', in J. Zubin (ed.) *Field Studies in the Mental Disorders*, pp. 73–93 (New York: Grune & Stratton).
Samuels, L. (1978) Sex differences in concordance rates for schizophrenia. *Schizophrenia Bulletin*, 4, 14–15.

Savodnik, I. (1978) 'The manifest and scientific images', in: J.C. Shershow (ed.) *Schizophrenia: Science and Practice*, pp. 21–47 (Cambridge, MA: Harvard University Press).

Schacht, T. and Nathan, P.E. (1977) But is it good for the psychologists? Appraisal and status of DSM-III. *American Psychologist*, 32, 1017–25.

Scheuer, K. (1957) A town called Hollywood. *Los Angeles Times*, Mar 17, p. 2

Schindler, R. (1957) 'Über die grundsätzliche stellung der psychotherapie bei psychosen', in: C. Muller and G. Benedetti (eds) *Symposium International sur la Psychothérapie de la Schizophrénie*, pp. 147–62 (Basel: Karger).

Schneider, K. (1918) Schizophrene Kriegspsychosen. *Zeitschrift für die gesamte Neurologie und Psychiatrie*, 43, 420–9.

Schneider, K. (1974) 'Primary and secondary symptoms in schizophrenia', in: S.R. Hirsch and M. Shepherd (eds) *Themes and Variations in European Psychiatry: An Anthology*, pp. 40–4 (Bristol: John Wright).

Schulz, S.C., Conley, R.R., Kahn, E.M. and Alexander, J. (1989) 'Nonresponders to neuroleptics: A distinct subtype', in: S.C. Schulz and C.A. Tamminga (eds) *Schizophrenia: Scientific Progress*, pp. 341–51 (Oxford: Oxford University Press).

Sechehaye, M.A. (1950) *Journal d'une Schizophrène* (Paris: Presse Universitaires de France).

Sedler, J. (1985) The legacy of Ewald Hecker: A new translation of Die Hebephrenie (M. Schoelly, trans.) *American Journal of Psychiatry*, 142, 1265–71.

Sérieux, P. (1913) Chaslin. *Revue générale des sciences pures et appliquées*, 24, 593–4.

Shakow, D., Rodnick, E.H. and Lebeaux, T. (1945) A psychological study of a schizophrenic: Exemplification of a method. *The Journal of Abnormal and Social Psychology*, 40, 154–74.

Shakow, D. (1977/2002) 'Clinical psychology seen some 50 years later', in: W. Pickren and D. Dewsbury (eds) *Evolving Perspectives on the History of Psychology*, pp. 433–51 (Washington, DC: American Psychological Association).

Shamis, S. (1976) Linguistic relativity and the diagnosis of schizophrenia. *Schizophrenia Bulletin*, 2, 503–4.

Shastry, B.S. (1999) Recent developments in the genetics of schizophrenia. *Neurogenetics*, 2, 149–54.

Shields, J. (1978) 'Genetics', in: J.K. Wing (ed.) *Schizophrenia: Towards a New Synthesis*, pp. 53–89 (London: Academic Press).

Shorter, E. (2006) Hysteria and catatonia as motor disorders in historical context. *History of Psychiatry*, 17, 461–8.

Silveira, A. (1999) Cerebral systems in the pathogenesis of endogenous psychoses. *Arquivos de Neuro-Psiuiatria*, 57, 1046–56.

Singer, H.D. (1917) Suggestions for a scheme of graphic representation of personality and psychosis. *The Journal of Abnormal Psychology*, 12, 114–29.

Smith, R. (1988) Does the history of psychology have a subject? *History of the Human Sciences*, 1, 147–77.

Smyth, B. (2011) Foucault and Binswanger. *Philosophy Today*, 55, 92–101.

Spitzer, R.L. (1982) Letter to the editor. *Schizophrenia Bulletin*, 8, 592.

Statt, D. (1981) *A Dictionary of Human Behaviour* (London: Harper & Row).

Stevenson, R.L. (1886/1890). *L'etrange cas du Dr. Jekyll et Mr Hyde* [*The Strange Case of Dr. Jekyll and Mr. Hyde*] (B.J. Lowe, trans.) (Paris: E. Plon, Nourrit et Cie).

Stockings, G.T. (1945) Schizophrenia in military psychiatric practice. *Journal of Mental Science*, 91, 110–12.

Stoddart. W. (1898) Certain physical signs in melancholia. *Journal of Mental Science [British Journal of Psychiatry]*, 44, 247–59.

Strauss, J.S., Bowers, M., Downey, T.W., Fleck, S., Jackson, S. and Levin, I. (1977) *The Psychotherapy of Schizophrenia* (New York: Jason Aronson).

Strecker, E.A. (1917) Certain of the clinical aspects of 'Late Katatonia' with a report of cases. *American Journal of Psychiatry*, 74, 185–220.

Strömgren, E. (1961) 'Defining the unit of study in field investigations in the mental disorders', in: J. Zubin. (ed.) *Field Studies in the Mental Disorders*, pp. 173–211 (New York: Grune & Stratton).

Sullivan, H.S. (1929) Research in schizophrenia. *American Journal of Psychiatry*, 9, 553–67.

Sullivan, H.S. (1964) *The Fusion of Psychiatry and Social Science* (New York: Vail-Ballou Press).

Szasz, T.S. (1960/1974) *The Myth of Mental Illness. Foundations of a Theory of Personal* Conduct, rev. ed. (New York: Harper Collins).

Tesson, E. (1952) La personne humaine et la chirurgie. Études, 275, 220–8.

Thakurdas, H. and Thakadurdas, L. (1979) *Dictionary of Psychology* (Lancaster: MTP Press).

The World Fellowship for Schizophrenia and Allied Disorders (2005) Schizophrenia, the myths, the signs, the statistics. Pamphlet 18. Available at: http://www.world-schizophrenia.org/publications/18-myths.html (accessed 25 August 2005).

Thornicroft, G. (1986) Contemporary Psychiatry in China. *British Medical Journal*, 292, 813–15.

Torey, E.F. (2001) *Surviving Schizophrenia*, 4th ed. (New York: Harper Collins).

Travis, L.E. (1924). Suggestibility and negativism as measured by auditory threshold during reverie. *Journal of Abnormal and Social Psychology*, 18, 350–68.

Trick, K.L. (1967). Schizophrenia in a male pseudohermaphrodite. *British Medical Journal*, 3, 783–4.

Urfer, A. (2001) Phenomenology and psychopathology of schizophrenia. The views of Eugene Minkowski. *Philosophy, Psychiatry, & Psychology*, 8, 279–89.

Van Nuys, K. (1953) Evaluating the pathological in phrophetic experience (particularly in Ezekiel). *Journal of Bible and Religion*, 21, 244–51.

Van Os, J., Galdos, P., Lewis, G, Bourgeois, M. and Mann, A. (1993) Schizophrenia sans frontieres: concepts of schizophrenia among French and British psychiatrists. *British Medical Journal*, 307, 489–92.

Vendée. Conseil général (1927) Rapports du préfet. La Roche-Sur-Yon: Georges Grillard. Available at: http://catalogue.bnf.fr/ark:/12148/cb34438121d (accessed 6 February 2015).

Voivenel, P. (1929) Science Médicales. *Mercure de France*, 212, 185–92.

Von Bueltzingsloewen, I. (2008) 'Une étape dans la mise en question du modèle asilaire? La famine dans les hôpitaux psychiatriques français sous l'occupation', in: J. Missa (ed.) *Les cahiers du centre Georges-Canguilhem. No. 2. Les maladies mentales*, pp. 47–61 (Paris: Presses Universitaires de France).

Vygotsky, L.S. (1993) *The Collected Works of L. S. Vygotsky Vol. 2. The Fundamentals of Defectology (Abnormal Psychology and Learning Disabilities)*, R. W. Rieber and A.S. Carton (eds.) (New York: Plenum Press).

Walton, H. (1985). *Dictionary of Psychiatry* (London: Blackwell Scientific Publications).

Warner, R. (1994) *Recovery from Schizophrenia: Psychiatry and Political Economy*, 2nd ed. (London: Routledge).

Watson, J.B. (1998). *Behaviorism* (Rutgers, NJ: Transaction Publishers).

Watson, S. (1998) *The Birth of the Beat Generation: Visionaries, Rebels, and Hipsters,1944–1960* (New York: Pantheo).

White, M. A. (1949). A study of schizophrenic language. *The Journal of Abnormal and Social Psychology*, 44, 61–74.

Wilkening, H.E. (1973) *The Psychology Almanac: A Handbook for Students* (Belmont, CA: Brooks/Cole Publishing Company).

Wittgenstein, L. (2001) *Philosophical Investigations* (Oxford: Blackwell Publishing).

Woodard, J.W. (1936). A new classification of culture and a restatement of the culture lag theory. *American Sociological Review*, 1, 89–102.

World Health Organization (1967) *International Classification of Diseases*, 8th ed. (Geneva: World Health Organization).

Young, P. (1988) *Schizophrenia* (London: Chelsea House Publications).

Zubin, J. (1987) Statistics in psychiatry. *Statistical Science*, 2, 121–5.

References

Abou-Saleh, M. T. (1987) Is schizophrenia a neurodevelopmental disorder? *British Medical Journal (Clinical Research Edition)*, 295, 1278.

Abraham, K. (1908/1927) 'The psychosexual differences between hysteria and dementia Praecox', in: Bryan, D. and Stracky, A. (eds) *Selected Papers of Karl Abraham*, pp. 64–79 (London: Hogarth).

Acocella, J. (1999) *The Diary of Vaslav Nijinsky* (New York: Farrar, Strauss and Giroux).

Allard, P. (1941). Une nouvelle maladie: le dinguaullisme. *Le Matin*, 1–4.

Allen, C. (1936) The phantasy of murder. *The Lancet*, 227, 1441.

Aly, G. (1994) 'Medicine Against the Useless', in: Aly, G., Chroust, P. and Pross, C. (eds) *Cleansing the Fatherland: Nazi Medicine and Racial Hygiene*, pp. 22–237 (Baltimore, MD: Johns Hopkins University Press).

American Medical Association (1937) *Standard Classified Nomenclature of Disease* (Chicago, IL: American Medical Association).

American Medico-Psychological Association and National Committee for Mental Hygiene (1918) *Statistical Manual for the Use of Institutions for the Insane* (New York: American Medico-Psychological Association and National Committee for Mental Hygiene).

American Psychiatric Association (1952) *The Diagnostic and Statistical Manual of Mental Disorders* (Washington, DC: American Psychiatric Association).

American Psychiatric Association (1968) *The Diagnostic and Statistical Manual of Mental Disorders*, 2nd ed. (Washington, DC: American Psychiatric Association).

American Psychiatric Association (1980) *The Diagnostic and Statistical Manual of Mental Disorders*, 3rd ed. (Washington, DC: American Psychiatric Association).

American Psychiatric Association (1987) *The Diagnostic and Statistical Manual of Mental Disorders*, 3rd rev. ed. (Washington, DC: American Psychiatric Association).

American Psychiatric Association (2004) *The Diagnostic and Statistical Manual of Mental Disorders DSM-IV*-TR, 4th ed. (Arlington, VA: American Psychiatric Association).

Andreason, N.C. (1994a) 'Characteristic symptoms of schizophrenia', in: Widiger, T.A., Frances, A.J., Pincus, H.A., First, M.B., Ross, R. and Davis, W. (eds) *DSM-IV Sourcebook*, Vol. 1, pp. 351–80 (Washington, DC: American Psychiatric Association).

Andreason, N.C. (1994b) 'Schizophrenia and other psychotic disorders', in: Widiger, T.A., Frances, A.J., Pincus, H.A., First, M.B., Ross, R. and Davis, W. (eds), Vol. 1, pp. 343–51 (Washington, DC: American Psychiatric Association).

Andreason N.C. (1999) A unitary model of schizophrenia: Bleuler's fragmented phrene as schizencephaly. *Archives of General Psychiatry*, 56, 781–7.

Andreason, N.C., Hoffman, R.E. and Grove, W.M. (1984) 'Mapping abnormalities in language and cognition', in: Alpert, M. (ed.) *Controversies in Schizophrenia*, pp. 199–226 (London: Guilford Press).

Angel, K., Jones, E. and Neve, M. (2003) *European Psychiatry on the Eve of War: Aubrey Lewis, the Maudsley Hospital and the Rockefeller Foundation in the 1930s. Medical History, Supplement No. 22* (London: The Wellcome Trust for the History of Medicine at UCL).

Annitto, W.J. (1977) On the existence of madness. *Schizophrenia Bulletin*, 3, 520–1.

Anon. (1873) Ordentliche versammlung des psychiatrischen vereines zu Berlin am 15. Marz 1872. *Allgemeine Zeitschrift für Psychiatrie und Psychisch-Gerichtliche Medizin*, 29, 455.

Anon. (1892) Irish meeting. *Journal of Mental Science [British Journal of Psychiatry]*, 38, 639–40.

Anon. (1902) British Medical Association. *Journal of Mental Science [British Journal of Psychiatry]*, 48, 806–8.

Anon. (1923). Annual meeting at Portsmouth. Prognosis of psychosis during adolescence. *The Lancet* 202, 457–63.

Anon. (1927). A l'université. *Gazette de Lausanne*, Mar 3, p. 2.

Anon. (1930) Budapest. The nervous and mental indications for artificial abortion. *The Lancet*, 215, 420–1.

Anon. (1932) Paul Bazan, le refle de l'ange. *Revue des Lectures*, 6, 612–13.

Anon. (1933) Eugenics in Germany. The question of voluntary sterilisation. *The Lancet*, 221, 1203–4.

Anon. (1934) Artificial pyrexia in schizophrenia. *The Lancet*, 223, 985.

Anon. (1936) Insulin in schizophrenia. *The Lancet*, 227, 1418–19.

Anon. (1938) Treatment of schizophrenia. *The Lancet*, 232, 1184–5.

Anon. (1947) Schizophrenia strikes 1 or 2 percent of the population of the United States. *Science*, 106, 12.

Anon. (1951) Dementia praecox or the group of schizophrenias. *Journal of Consulting Psychology*, 15, 167.

Anon. (1956) Les schizophrènes Soviétiques soignés en Allemange ne sont pas descobayes. *Gazette de Lausanne*, May 24, p. 8.

Anon. (1957a) Psychiatrists discuss schizophenia. *The British Medical Journal*, 2, 755–6.

Anon. (1957b) Insanity drug's effects stopped by treatment. *The Science News-Letter*, 72, 344–5.

Anon. (1957c). L'abus des calmants. *Journal de Genève*, Jul 19, p. 1.

Anon. (1958). Les médicaments dits tranquillisants, *Journal de Genève*, Nov 1, p. 4.

Anon. (1959). Le Dr Fisher et le Japon. *Journal de Genève*, May 21, p. 4.

Anon. (1965) Changes in schizophrenia. *British Medical Journal*, 1, 141–2.

Anon. (1970a) Phenomenology in perspective. Is there schizophrenia? *Schizophrenia Bulletin*, 1, 11–14.

Anon. (1970b) NIMH activities. *Schizophrenia Bulletin*, 1, 75–6.

Anon. (1970c) At issue: The psychosocial treatment of schizophrenia. *Schizophrenia Bulletin*, 1, 4–5.

Anon. (1970d) NIMH activites. *Schizophrenia Bulletin*, 1, 79–83.

Anon. (1970e) NIMH activities: Conference on the implications of research findings for treatment and training. *Schizophrenia Bulletin*, 1, 43–5.

Anon. (1972a) Does schizophrenia get its start early in life? *Science News*, 102, 263.

Anon. (1972b) Chronic schizophrenia: An iatrogenic condition? *Schizophrenia Bulletin*, 1, 4.

Anon. (1974a) United Kingdom Cross-National Project (1974). The diagnosis and psychopathology of schizophrenia in New York and London. *Schizophrenia Bulletin*, 1, 80–102.

Anon. (1974b) Unity and diversity in schizophrenia. *British Medical Journal*, 4, 673–4.

Anon. (1977a) On being diagnosed schizophrenic. *Schizophrenia Bulletin*, 3, 4.

Anon (1978a) 'Genetic depression' and 'viral schizophrenia'. *Science News*, 114, 244.

Anon. (1978b) The 'schizophrenics' who aren't. *Science News*, 114, 230–1.

Anon. (1981) Abuse of psychiatry against dissenters. *Economic and Political Weekly*, 16, 185–8

Appelbaum, P.S. (2001) Abuses of law and psychiatry in China. *Law and Psychiatry*, 52, 1297–8.

Arieti, S. (1955) *Interpretation of Schizophrenia* (New York: Robert Brunner).

Arieti, S. (1968) *Schizophrenia. Encyclopedia Britannica*, Vol. 19, p. 1162 (London: William Benton).

Arieti, S. (1974a) *American Handbook of Psychiatry*, Vol. 3 (New York: Basic Books).

Arieti, S. (1974b) *Interpretation of Schizophrenia*, 2nd ed. (London: Crosby Lockwood Staples).

Arnold, H.M. (1976) Working with schizophrenic patients. Four A's: A guide to one-to-one relationships. *American Journal of Nursing*, 76, 941–3.

Arthur, R.J. (1973) Social psychiatry: An overview. *American Journal of Psychiatry*, 130, 841–9.

Ban, T.A. (2004) Neuropsychopharmacology and the genetics of schizophrenia. A history of the diagnosis of schizophrenia. *Progress in Neuro-Psychopharmacology and Biological Psychiatry*, 28, 753–62.

Bannister, D. (1971) The logical requirements of research into schizophrenia. *Schizophrenia Bulletin*, 1, 72–7.

Barnes, M. and Berke, J. (1973) *Two Accounts of a Journey Through Madness* (New York: Ballentine Books).

Bartko, J. and Carpenter, W.T. (1977) The masks of insanity: a graphics tool for the summarization of psychiatric data. *Schizophrenia Bulletin*, 3, 632–8.

Basaglia, F. (1953) Il mondo dell'«incomprensibile» schizofrenico attraverso la Daseinsanalyse [The world of the schizophrenic incomprehensible through Daseinsanalyse; presentation of a case]. *Giornale Psichiatria Neuropatol*, 81, 471–99.

Basaglia, F. (1968/2013) *L'istituzione Negata* (Milan: Baldini and Castoldi).

Baynes, H.G. (1940) *Mythology of the Soul* (London: Routledge and Kegan Paul).

Beck, S.J. (1954) *The Six Schizophrenias* (New York: The American Orthopsychiatric Association).

Becker, H. (1936) A new classification of culture and a restatement of the culture lag theory: Discussion. *American Sociological Review*, 1, 102–4.

Beckett, P. (1971) 'Clinical presentation and classification of schizophrenia', in: Tourney, G. and Gottlieb, J.S. (eds) *Lafayette Clinical Studies on Schizophrenia*, pp. 69–95 (Detroit, MI: Wayne State University Press).

Beel, C. (1975) Family and social management of schizophrenia. *Schizophrenia Bulletin*, 1, 97–118.

Beers, C. (1908) *A Mind that Found Itself* (New York: Longmans Green and Company).

Behr, A. (1891) *Die frage der 'Katatonie' oder des irreseins mit spanning* (Riga: W. F Häcker).

Beigel, H.G. (1971) *Dictionary of Psychology and Related Fields* (London: George G. Harrap).

Beja, M. (1978) Bartleby and schizophenia. *The Massachusetts Review*, 19, 555–68.

Bellak, L. (1947/1952) *Dementia Praecox. The Past Decade's Work and Present Status: A Review and Evaluation* (New York: Grune & Stratton).

Bellak, L. (1948) *Dementia Praecox. The Past Decade's Work and Present Status: A Review and Evaluation* (New York: Grune & Stratton).

Bellak, L. (1961) *Contemporary European Psychiatry* (New York: Grove Press).

Bellak, L. (1975) Intercultural studies in search of a disease. *Schizophrenia Bulletin*, 1, 6–9.

Bellak, L. (1977) A drug-free week after admission. *Schizophrenia Bulletin*, 3, 342–4.

Bellak, L. and Loeb, L. (1969) *The Schizophrenic Syndrome* (London: Grune and Stratton).

Bender, L., Cobrinik, L., Faretra, G. and Sankar, S. (1966) 'The treatment of childhood schizophrenia with LSD and UML', in: Rinkel, M. (ed.) *Biological Treatment of Mental Illness*, pp. 463–91 (New York: New York Academy of Medicine).

Bennet, I.F. (1956) 'Cooperative VA study of Chemotherapy in Psychiatry. Project No. 1', in: Gerard, R.W. and Cole, J.O. (ed.) *Psychopharmacology. Problems in Evaluation*, pp. 412–20 (Washington, DC: National Academy of Sciences–National Research Council).

Bentall, R. (2003) *Madness Explained. Psychosis and Human Nature* (London: Penguin Books).

Berke, J. (1977) *I Haven't had to go Mad Here* (London: Pelican Books).

Berrios, G. (1996) *The History of Mental Symptoms* (Cambridge: Cambridge University Press).

Berrios, G. and Kraam, A. (2002) Classic text no. 50. On heboidophrenia. *History of Psychiatry*, 13, 197–208.

Berrios, G.E., Luque, R. and Villagrán, J.M. (2003) Schizophrenia: A conceptual history. *International Journal of Psychology and Psychological Therapy*, 3, 111–40.

Biddle, E. (1949) The nurse and 'spontaneous' recovery in schizophenia. *The American Journal of Nursing*, 49, 371–2.

Binswanger, L. (1957) *Schizophrenie* [Schizophrenia] (Pfullingen: Neske Verlag).

Birnbaum, K. (1931) *Kriminal psychopathologie und psychobiologische verbrecherkunde* (Berlin: Springer).

Blashfield, R.K. (1982) Feighner et al., invisible colleges, and the Matthew effect. *Schizophrenia Bulletin*, 8, 1–6.

Bleckwenn W.J. (1930) Production of sleep and rest in psychotic cases. *Archives of Neurology and Psychiatry*, 24, 365–75.

Bleuler, E. (1906/1969) 'Consciousness and association (M.D. Eder, trans.)', in: Jung, C.G. (ed.) *Studies in Word-association*, pp. 266–296 (London: Routledge and Keegan Paul).

Bleuler, E. (1908) Die prognose der dementia praecox – schizophreniegruppe [The prognosis of the dementia praecox – schizophrenia group]. *Allgemeine Zeitschrift für Psychiatric und Psychisch-Gerichtliche Medizin*, 65, 436–64.

Bleuler, E. (1911) *Dementia praecox oder gruppe der schizophrenien* (Leipzig: Franz Deuticke).

Bleuler, E. (1911/1950) *Dementia Praecox or the Group of Schizophrenias* (J. Zinkin, trans.) (New York: International Universities Press).

Bleuler, E. (1911/1952) *Dementia Praecox or the Group of Schizophrenias*, 2nd ed. (J. Zinkin, trans.) (New York: International Universities Press).

Bleuler, E. (1912) *The Theory of Schizophrenic Negativism* (W.A. White, trans.) (New York: Journal of Nervous and Mental Diseases Publishing Company).

Bleuler, E. (1912/1951) 'Autistic thinking', in: Rapaport, D. (ed.) *Organisation and Pathology of Thought. Selected Sources*, pp. 397–437 (New York: Columbia University Press).

Bleuler, E. (1916/1924) *Textbook of Psychiatry* (A.A. Brill, trans.) (New York: Macmillan).

Bleuler, E. (1926) *La schizophrénie* [*Schizophrenia*] (Paris: Masson).

Bleuler, E. (1930a) *Lehrbuch der psychiatrie. Fünfte Auflage* [*Textbook of Psychiatry*, 5th ed.] (Berlin: Julius Springer).

Bleuler, E. (1930b) The physiogenic and psychogenic in schizophrenia. *American Journal of Psychiatry*, 10, 203–11.

Bleuler, M. (1966) *Conception of Schizophrenia Within the Last Fifty Years and Today* (New York: The International Science Press).

Bleuler, M. (1972/1978) *The Schizophrenic Disorders. Long-term Patient and Family Studies* (New Haven, CT: Yale University Press). [Originally published as *Die schizophrenen Geistesstörungen im Lichte langjähriger Kranken und Familiengeschichten* (Stuttgart: Georg Thieme Verlag)]

Bleuler, E. and Claude, H. (1926/2001) *La schizophrénie en débat* [*Schizophrenia in Debate*] (Chazaud, J., ed.) (Paris: L'Harmattan).

Bloch, S. and Reddaway, P. (1977) *Russia's Political Hospitals* (London: Victor Gollancz).

Bobon, J. (1952) Schizoparalexia and schizoparagraphia. *Acta Neuroligica et Psychiatrica Belgica*, 52, 286–95.

Bobon, J. (1967) Schizophasia and schizoparaphasia. Coherence of spontaneous drawing; importance of tests of anagrams, reversal and mirror inversion of words. *Acta Neuroligica et Psychiatrica Belgica*, 67, 924–38.

Bordieu, P. (1990) *The Logic of Practice* (Cambridge: Polity Press).

Borges, J.L. (1971/1979) *The Book of Sand* (N. Giovanni, trans.) (London: Penguin Books).

Boring, E. Bridgman, P., Feigl, H., Israel, H., Pratt, C. and Skinner, B. (1945) Symposium on operationism. *The Psychological Review*, 52, 241–94.

Bornstein, M. (1917) Über einen eigenartigen typus der psychischen spaltung 'Schizothymia reactiva'. *Zeitschrift für die gesamte Neurologie und Psychiatrie*, 36, 86–145.

Bourgeois, M.L. (1999) *Les schizophrénies* [*The Schizophrenias*] (Paris: Presses Universitaires de France).

Bower, B. (1985) Identifying the early warning signs of schizophrenia. *Science News*, 125, 311.

Bowman, K.M. and Kasanin, J. (1933) Constitutional schizophrenia. *American Journal of Psychiatry*, 8, 645–57.

Boyle, M. (1990) *Schizophrenia. A Scientific Delusion?* (London: Routledge).

Boyle, M. (2001) *Schizophrenia. A Scientific Delusion?*, 2nd ed. (London: Routledge).

Braconnier, A. (2006) *Les âges de la vie* [*The Ages of Life*] (Paris: Masson).

Bradley, C. (1941) *Childhood Schizophrenia* (New York: The Macmillan Company).

Braslow J.T. (1996) In the name of therapeutics: The practice of sterilization in a California state hospital. *Journal of the History of Medicine and Allied Sciences*, 51, 29–51.

Bridgman, P.W. (1927) *The Logic of Modern Physics* (New York: Macmillan).
Bridges, J.W. (1919) *An Outline of Abnormal Psychology* (Columbus, OH: R.G. Adams).
Brill, A.A. (1909) A case of schizophrenia (dementia praecox). *American Journal of Insanity*, 66, 53–70.
Brill, A.A. (1925/1928) 'The neuroses and psychoses', in: Dana, C.L., Davis, T.K., Jelliffe, S.E., Riley, H.A., Tilney, F. and Timme, W. (eds) *Schizophrenia [Dementia Praecox]; An Investigation by the Association for Research in Nervous and Mental Diseases*, pp. 31–43 (New York: Paul B. Hoeber).
Brill, A.A. (1949) *Lectures on Psychoanalytic Psychiatry* (London: John Lehmann).
Brill, N., Corcoran, J.F., Eiduson, S., Judd, L., May, P., Stoller, R. and Walter, R. (1969) The UCLA interdepartmental conference: schizophrenia. *Annals of Internal Medicine*, 70, 107–25.
Brissot, M. (1913) La démence précoce ou Schizophrenie d'après la conception de Bleuler, par M. Trénel. *Revue de Psychiatrie et de Psychologie Expérimentale*, 1, 307–8.
Brock, A. (ed.) (2009) *Internationalising the History of Psychology* (New York: NYU Press).
Brockington, I. (1985) 'Diagnosis of schizophrenia and schizoaffective psychoses', in: Bradley, P.B. and Hirsch, S.R. (eds) *The Psychopharmacology and Treatment of Schizophrenia*, pp. 166–200 (Oxford: Oxford University Press).
Brosius, C. (1877) Die katatonie. Eine psychiatrische skizze. *Allgemeine Zeitschrift fürPsychiatrie*, 33, 770–802.
Brown, J. (1880) *Iets over katatonie* (Voorburg: W.H.C. Susan).
Brown, S.S., Kopin, I.J., Kramer, M., Mosher, L., Pollin, W.N., Salmoiraghi, G.C. and Torrey, E.F. (1974) The 1972 NIMH mission to assess schizophrenic research in the U.S.S.R. *Schizophrenia Bulletin*, 1, 148–153.
Bruce, L. and Peebles, A. (1903) Clinical and experimental observations on catatonia. *Journal of Mental Science*, 49, 614–28.
Bruno, F.J. (1986) *Dictionary of Key Words in Psychology* (London: Routledge & KeganPaul).
Buchsbaum, M. (1977) Psychophysiology and schizophrenia. *Schizophrenia Bulletin*, 3, 7–14.
Buchsbaum, M.S. and Haier, R.J. (1978) Biological homogeneity, symptom heterogeneity and the diagnosis of schizophrenia. *Schizophrenia Bulletin*, 4, 473–5.
Bührmann, M.V. (1966) Childhood schizophrenia. *South African Medical Journal*, 40, 920–3.
Burston, D. (1996) *The Wing of Madness. The Life and Work of R.D. Laing* (Cambridge, MA: Harvard University Press).
Burston, D. (1998) Family, phenomenology and schizophrenia in R.D. Laing. Available at: http://laingsociety.org/colloquia/psychotherapy/fps4.htm (accessed 25 August 2015).
Burston, D. (2001) R.D. Laing and the politics of diagnosis. Available at: http://laingsociety.org/colloquia/polofdiagnosis/rdlpoldx.htm (accessed 25 August 2015).
Buss, A.H. (1966) *Psychopathology* (New York: John Wiley).
Calanca, A. (1974) *La schizophrénie simple. Histoire et évolution d'un diagnostic* (unpublished PhD dissertation), Université de Lausanne.
Cameron, D.E. (1935) *Objective and Experimental Psychiatry* (New York: Macmillian).

Campbell, C. (1929/1934) 'A discussion of some errors in the diagnosis of schizophrenia', in: C. Campbell and E. Wilson (eds) *Statistical Studies from the Boston Psychopathic Hospital, 1925–1934*, pp. 41–51 (Concord, NH: Rumford Press).

Campbell, C. (1943) Clinical studies in schizophrenia: A follow-up study of a small group of cases of deterioration with few special trends (schizophrenic 'surrender'). *American Journal of Psychiatry*, 99, 475–83.

Cantor, M.B. (1968) Problems in diagnosing and prognosing with occult schizophreniapatients. *American Journal of Psychoanalysis*, 39, 46–7.

Capote, T. (1966/2005) *In Cold Blood* (London: Penguin).

Cardwell, M. (1986) *Dictionary of Psychology* (Chicago, IL: Fitzroy Dearborn Publishers).

Carlson, N.R., Martin, G.N. and Buskist, W. (2004) *Psychology*, 2nd European ed. (London: Pearson Education).

Carpenter, W.T. and Buchanan, R.W. (1995) 'Schizophrenia: Introduction and overview', in: H. Kaplan and B. Sadock (eds) *Comprehensive Textbook of Psychiatry*, 6th ed., Vol. 1, p. 889 (London: Williams & Wilkins).

Ceccherinie-Nelli, A. and Crow, T.J. (2003) Disintegration of the components of language as the path to a revision of Bleuler's and Schneider's concepts of schizophrenia: Linguistic disturbances compared with first-rank symptoms in acute psychosis. *British Journal of Psychiatry*, 182, 233–40.

Chapman, L.J. and Chapman, J.P. (1973) *Disordered Thought in Schizophrenia* (Upper Saddle River, NJ: Prentice Hall).

Chiswick, D. (1981) Matricide. *British Medical Journal (Clinical Research Edition)*, 283, 1279–80.

Claridge, G. (1978) Animal models of schizophrenia: the case for LSD-25. *Schizophrenia Bulletin*, 4, 186–209.

Clarke, J.W. (1981) American assassins: an alternative typology. *British Journal of Political Science*, 11, 81–104.

Claude, H. (1926) *Démence précoce et schizophrenie* [*Dementia Praecox and Schizophrenia*] (Paris: Masson).

Claude, H. (1937) Relation of hysteria to schizophrenia. *Annales Medico-Psychologiques*, 95, 1–14.

Cleland, C.C. (1975) Schizophrenic's role in diagnosis. *Schizophrenia Bulletin*, 1, 15–16.

Cline, S. (2002) *Zelda Fitzgerald: Her Voice in Paradise* (London: John Murray).

Cocks, G. (1997) *Psychotherapy in the Third Reich. The Goring Institute*, 2nd ed. (London: Transaction Publishers).

Cole, R.H. (1913) *Mental Diseases. A Text-book of Psychiatry for Medical Students and Practitioners* (New York: Wood).

Cooper, D. (1967) *Psychiatry and Anti-psychiatry* (London: Paladin).

Cooper, D. (1970) *Psychiatry and Anti-psychiatry* (London: Paladin).

Cooper, J.E., Kendell, R.E., Gurland, B.J., Sharpe, L. Copeland, J.R.M. and Simon, R. (1972) *Psychiatric Diagnosis in New York and London* (London: Oxford University Press).

Cooper, R. (2007) *Psychiatry and Philosophy of Science* (Stocksfield: Acumen).

Coppock, V. and Hopton, J. (2000) *Critical Perspectives on Mental Health* (London: Routledge).

Coriat, I. (1914) Recent trends in the psychopathology of dementia praecox. *American Journal of Psychiatry*, 70, 669–82.
Coriat, I. (1917) Some statistical results of the psychoanalytic treatment of the psycho-neuroses. *Psychoanalytic Review*, 4, 209–17.
Corning, W.C. and Steffy, R.A. (1979) Taximetric strategies applied to psychiatric classification. *Schizophrenia Bulletin*, 5, 294–305.
Costello, C.G. (1970). *Symptoms of Psychopathology* (New York: John Wiley & Sons).
Crookshank, F.G. (1930) Types of personality, with special reference to individual psychology.*The Lancet*, 215, 546–8.
Crow, T.J. (1980) Molecular pathology of schizophrenia: More than one disease process? *British Medical Journal*, 280, 66–8.
Curran, W.J. (1974) Ethical and legal considerations in high risk studies of schizophrenia. *Schizophrenia Bulletin*, 1, 74–92.
Curran, D. and Guttmann, E. (1944) *Psychological Medicine. A Short Introduction to Psychiatry*, 1st ed. (London: E. & S. Livingstone).
Curran, D. and Guttman, E. (1946) *Psychological Medicine*, 2nd ed. (Edinburgh: E. & S. Livingstone).
Curran, D. and Partridge, M. (1955) *Psychological Medicine: A Short Introduction to Psychiatry*, 4th ed. (London: E. & S. Livingstone).
Dagand, H. (1936) Un cas de catatonie intermittent pure suivi depuis vingt quatre ans. *Encéphale*, 31, 293–9.
Danziger, K. (1993) Psychological objects, practice, and history. *Annals of Theoretical Psychology*, 8, 15–46.
Darwin. C. (1871/1981) *The Descent of Man, and Selection in Relation to Sex* (Princeton, NJ: Princeton University Press).
Daston, L. (2000) *Biographies of Scientific Objects* (Chicago, IL: University of Chicago Press).
Davis, A.E., Dinitz, S. and Pasamanick, B. (1974) *Schizophrenics in the New Custodial Community* (Columbus, OH: Ohio State University Press.
Davis, D. (1950) A case of schizosis with dual personality. *Journal of Mental Science*, 96, 1008–14.
Decker, H. (1986) 'The lure of nonmaterialism in materialist Europe: Investigations in dissociative phenomena, 1880–1915', in: Quen, J.M. (ed.) *Split Minds/Split Brains. Historical and Current Perspectives* (New York: New York University Press).
Decker, H. (2013) *The Making of DSM-III. A Diagnostic Manual's Conquest of American Psychiatry* (Oxford: Oxford University Press).
De Fleury (1924) Une classification des malades mentales. *Bulletin de l'Académie nationale de médecine*, 91, 45–56.
Delay, J. and Deniker, P. (1955) Hibernothérapies et cures neuroleptiques en psychiatrie. *Bulletin de L'académie Nationale de Médicine*, 139, 145–6.
Deleuze, G. and Guattari, F. (2004) *A Thousand Plateaus: Capitalism and Schizophrenia* (London: Continuum).
Despert, J.L. (1938) Schizophrenia in children. *Psychiatric Quarterly*, 12, 336–71.
Dhunjibhoy, J.E. and Bomb, M.B. (1931) Sulpher injections in the psychoses. *The Lancet*, 218, 1407–8.
Dialetics of Liberation (2012). 1967 dialectics participants. Available at: http://www.dialecticsofliberation.com/ (accessed 1 January 2015).
Diefendorf, A. (1902/1908) *Clinical Psychiatry* (London: MacMillan).

Diefendorf, A. (1918) *Clinical Psychiatry* (New York: Macmillian).
Dixon, B. (1970) Heretics are imporant. *Schizophrenia Bulletin*, 1, 5–6.
Dixon, B. (1989) Schizophrenia controversy. *British Medical Journal*, 298, 265.
Dora, I. (1963) Piqué sur le vif. *Journal de Genève*, May 18, p. 10.
Dorcus, R. M. and Shaffer, G.W. (1934) *Textbook of Abnormal Psychology* (London: George Allen and Unwin).
Dowbiggin, I. (1991) *Inheriting Madness. Professionalization and Psychiatric Knowledge in Nineteenth Century France* (Oakland, CA: University of California Press).
Dowbiggin, I.R. (1997) *Keeping America Sane. Psychiatry and Eugenics in the United Statesand Canada 1880–1940* (London: Cornell University Press).
Drapes, T. (1906) A note on psychiatric terminology. *The Journal of Mental Science*, 52, 75–84.
'Dreamy Schizophrene' (1968) *The Listener*, Aug 29, p. 260.
Duchet, A. (1975). *Difficulté et errerurs diagnostiques rencontrées dans le formes d'entrée de la schizophrénie* (unpublished PhD dissertation), Université de Besancon.
Durell J. and Katz, M. (1977) The changing clinical picture of schizophrenia. *Schizophrenia Bulletin*, 3, 528–31.
Durkheim, E. (1995) *The Elementary Forms of Religious Life* (K.E. Fields, trans.) (New York: The Free Press).
Eastman, I. (1945) *The World Almanac and Book of Facts for 1945* (New York: New York World-Telegram).
Elkins, W.F. (1986) William Lauron De Laurence and Jamacian Folk Religion. *Folklore*, 97, 215–18.
Ellard, J. (1977) Schizophrenia: changing concepts of a common psychic disability. *Modern Medicine*, 2, 13–17.
English, H.B. (1934) *A Student's Dictionary of Psychological Terms* (London: Harper & Brothers Publishers).
Engstrom, E. (2003) *Clinical Psychiatry in Imperial Germany* (New York: Cornell University Press).
Epstein, S. (1979) Natural healing processes of the mind: I. Acute schizophrenic disorganisation. *Schizophrenia Bulletin, National Institute of Mental Health*, 5, 313–21.
Eraso, Y. (2010) 'A burden to the state: the reception of the German "Active Therapy" in an Argentinean "Colony-Asylum" in the 1920s and 1930s', in: Ernst, W. and Mueller, T. (eds) *Transnational Psychiatries: Social and Cultural Histories of Psychiatry in Comparative Perspective, c. 1800–2000*, pp. 51–80 (Cambridge: Cambridge Scholars Publishing).
Evans, H. (1950) La pathologie de Louis Lambert: Balzac Aliéniste. *Revue D'HistoireLittérarire de la France*, 2, 249–304.
Evans, W.A. (1924). How to keep well. *Washington Post*, Jun 26, p. 10.
Ey, H. (1926/1969) *Dementia praecox ou groupe des schizophrenies. Par Bleuler* [*Dementia praecox or the group of schizophrenias. By Bleuler*] (Paris: Théraplix).
Ey, H. (1996) *Schizophrénie: études cliniques et psychopathologiques* (Le Plessis-Robinson: no publisher listed).
Ey, H., Arnaudies, J., Cazalis de Fontdouce, G. and Marciano, P. (1977) *La notion de schizophrénie: séminaire de Thuir, février-juin 75* (Paris: Desclée de Brower).

Eysenck, E. (1986) 'A critique of contemporary classification and diagnosis', in: Millon, T. and Klerman, G.L. (eds) *Contemporary Directions in Psychopathology. Towards the DSM-IV*, pp. 73–99 (New York: The Guilford Press).
Falstein, E. and Sutton, H. (1958) Childhood schizophrenia. *The American Journal of Nursing*, 58, 666–70.
Farber, M.L. (1938) A critique and investigation of Kretschmer's theory. *The Journal of Abnormal and Social Psychology*, 33, 398–404.
Faulkner, H.J. and Pruitt, V.D. (1988) *The Selected Correspondence of Karl A. Menninger. 1919–1945* (Columbia, MO: University of Missouri Press).
Feighner, J.P., Robins, E., Guze, S.B., Woodruff, R.A., Winokur, G. and Munoz, R. (1972) Diagnositic criteria for use in psychiatric research. *Archives of General Psychiatry*, 26, 57–63.
Feldstein, A. Hoagland, H. and Freeman, H. (1958) On the relationship of serotonin to schizophrenia. *Science*, 128, 358.
Fenton, W., Mosher, R. and Matthews, S. (1981) Diagnosis of schizophrenia. A critical review of current diagnositc systems. *Schizophrenia Bulletin*, 7, 452–75.
Féré, C. (1890) *Les épilepsies et les épileptiques* (Paris: F. Alcan).
Ferrier, I.N. (1985) Water intoxication in patients with psychiatric illness. *British Medical Journal (Clinical Research Edition)*, 291, 1594–6.
Fink, M. and Taylor, M.A. (2003) *Catatonia. A Clinicians Guide to Diagnosis and Treatment* (Cambridge: Cambridge University Press).
Fireside, H. (1979) *Soviet Psychoprisons* (New York: Norton).
Fish, F. (1960) Senile schizophrenia. *Journal of Mental Science*, 106, 938–46.
Fisher, V.E. (1937) *An Introduction to Abnormal Psychology,* revised ed. (New York: Macmillan).
Fleck, L. (1935/1979). *Genesis and Development of a Scientific Fact* (F. Bradley and T.J. Trenn, trans.) (Chicago, IL: University of Chicago Press).
Flew, A. (1977). Karl Kraus and the soul-doctors by Thomsas S. Szasz. *Reason*, 26–29. Available at: http://www.unz.org/Pub/Reason-1977nov-00026 (accessed 22 January 2015).
Folsom, C. (1886) *Mental Diseases* (Boston, MA: no publisher listed).
Forrest, D. (1976) Nonsense and sense in schizophrenic language. *Schizophrenia Bulletin*, 2, 286–301.
Foucault, M. (1954/1987) *Mental Illness and Psychology* (A. Sheridan, trans.) (Berkeley, CA: University of California Press).
Foucault, M. (1961/2006) *History of Madness* (London: Routledge).
Foucault, M. (1974) 'Noam Chomsky and Michel Foucault. Human nature: Justice versus power', in: F. Elders (ed.) *Reflexive Water. The Basic Concerns of Mankind*, pp. 133–99 (London: Souvenir Press).
Foucault, M. (1966/2006) *The Order of Things* (New York: Routledge).
Fowler, R.C. (1978) Remitting schizophrenia as a variant of affective disorder. *Schizophrenia Bulletin*, 4, 68–77.
Freeman, A. (1895) A case of Kahlbaum's so-called katatonia. *Journal of Nervous and Mental Disease*, 20, 814.
Freud, S. (1914/1981) 'On narcissism: An introduction', in: J. Strachey and A. Freud (eds) *The Standard Edition of the Complete Works of Sigmund Freud*, Vol. xiv, pp. 74–7 (London: The Hogart Press).
Fromm-Reichmann, F. (1948/1959) 'Notes on the development of treatment of schizophrenics by psychoanalytic psychotherapy', in: Bullard, D.M. (ed.)

Psychoanalysis and Psychotherapy: Selected Papers of Frieda Fromm-Reichmann, pp. 160–75 (Chicago, IL: The University of Chicago Press).

Furter, P. (1963). Jeunes gens en colère. *Journal de Genève*, Nov 30, p. 10.

Gabel, J. (1960) Le concept d'aliénation politique. *Revue française de sociologie*, 1, 454–464.

Gabel, J. (1997) *Ideologies and the Corruption of Thought* (Piscataway, NJ: Transaction Publishers).

Gair, D. (1972). A commentary. *Schizophrenia Bulletin*, 1, 55–61.

Garrabé, J. (2003) *La schizophrénie* [*Schizophrenia*] (Paris: Les Empêcheurs de Penser en Rond).

Gazzaniga, M.S., Bogen, J.E. and Sperry, R.W. (1962) Some functional effects of sectioning the cerebral commissures in man. *Proceedings of the National Academy of Science*, 48, 1765–9.

Geis, G. and Bienen, L.B. (1998) *Crimes of the Century* (Boston, MA: North Eastern University Press).

Gelinas, G. (1977) *Epistemologie du concept de schizophrénie* (unpublished PhD dissertation), Université du Qubec.

Gelman, S. (1999) *Medicating Schizophrenia: A History* (New Brunswick, NJ: Rutgers University Press).

Gerard, R.W. (1964) 'The nosology of schizophrenia: A co-operative study', in: R. Stanley (ed.) *Schizophrenia. Addresses, Essays, Lectures*, pp. 69–93 (New York: MSS Information).

Gillin, J.C., Kaplan, J. Stillman, R. and Wyatt, R.J. (1976) The psychedelic model of schizophrenia: the case of N,N-dimethultryptamine. *American Journal of Psychiatry*, 133, 203–8.

Gilman, S. (1985) The mad man as artist: medicine, history and degenerate art. *Journal of Contemporary History*, 20, 575–97.

Gilyarovsky, V.A. (1961) 'The Soviet Union', in: Bellak, L. (ed.) *Contemporary European Psychiatry*, pp. 283–327 (New York: Grove Press).

Ginguene, P.L. (1794) *La feuille villageoise: adressée, chaque semaine, à tous les villages de France, pour les instruire des loix des evénemens, des découvertes qui intétesesent tout citoyen* (Paris: De L'Imprimerie, rue Théatre du Peuple).

Gjessing, R. (1938) Disturbances of somatic function in catatonia with a periodic-course and their compensation. *Journal of Mental Science*, 84, 608–21.

Glueck, B. (1963) Early sexual experiences in schizophrenia. *Advances in Sex Research*, 1, 248–55.

Goffman, E. (1961) *Asylums: Essays on the Social Situation of Mental Patients and Other Inmates* (New York: First Anchor Books).

Goldman, H. (1977) Nonunitary disease hypothesis. *Schizophrenia Bulletin*, 3, 3.

Goldstein, G. (1978) Cognitive and perceptual differences between schizophrenics and organics. *Schizophrenia Bulletin*, 4, 160–85.

Goldstein, J. (1990) *Console and Classify (*Cambridge: Cambridge University Press).

Goodall, E. (1892) Observations upon 'Katatonia'. *The British Journal of Psychiatry*, 38, 227–33.

Goplerud, E. (1978) The diagnositic ambiguity of postpsychotic depression. *Schizophrenia Bulletin*, 4: 477–80.

Gottesman, I.I. (1991) *Schizophrenia Genesis: The Origins of Madness* (New York: W. H. Freeman).

Gottesman, II. and Shields, J. (1967) A polygenic theory of schizophrenia. *Proceedings of the National Academy of Sciences of the United States of America*, 58, 199–205.

Gottesman, I.I. and Shields, J. (1972) *Schizophrenia and Genetics: A Twin Study Vantage Point* (London: Academic Press).

Gottesman, I. and Shields, J. (1976) A critical review of recent adoption, twin, and family studies of schizophrenia: behavioural genetics perspectives. *Schizophrenia Bulletin*, 2, 360–401.

Grantt, W.H. (1953) Principles of nervous breakdown. Schizokinesis andautokinesis. *Annals of the New York Academy of Sciences*, 56, 143–63.

Gray, J.A. (1998) Integrating schizophrenia. *Schizophrenia Bulletin*, 24, 249–66.

Greenacre, P. (1918) The content of the schizophrenic characteristics occurring in affective disorders. *American Journal of Insanity*, 75, 197–202.

Greenberg, J.A. (1975) Schizophrenia: heretical notes on a diagnostic category. *Schizophrenia Bulletin*, 1, 10–14.

Greenberg, J. (1977) How accurate is psychiatry? *Science News*, 112, 28–9.

Greenberg, J. (1979) The schizophrenic brain: rewriting the chapter. *Science News*, 116, 26–7.

Gregory, R.L. (1987) *The Oxford Companion to the Mind* (Oxford: Oxford University Press).

Grinker, R.R. (1971) 'An essay on schizophrenia and science', in: Cancro, R. (ed.) *The Schizophrenic Syndrome*, pp. 53–93 (London: Butterworths).

Grinker, R. (1979) Diagnosis of borderlines: a discussion. *Schizophrenia Bulletin*, 5, 47–52.

Grob, G. (1991) Origins of DSM. A study in appearance and reality. *American Journal of Psychiatry*, 148, 421–31.

Grob, G. (1994) *The Mad Among Us: A History of the Care of America's Mentally Ill* (Cambridge, MA: USA).

Gross, G. (1938). Diabolical madmen made docile with new 'shock treatments'. *Washington Post,* Mar 7, p. 1.

Guggenheim, G.G. and Babigian, H.M. (1974) Diagnostic consistency in catatonic schizophrenia. *Schizophrenia Bulletin*, 1, 103–8.

Gunderson, J.G., Autry, J.H., Mosher, L. and Buchsbaum, S. (1974) Special report: schizophrenia 1974. *Schizophrenia Bulletin*, 1, 15–54.

Guntrip, H. (1969) *Schizoid Phenomena, Object Relations, and the Self* (New York: International Universities Press).

Hacking, I. (1995) *Rewriting the Soul. Multiple Personality and the Sciences of Memory* (Princeton, NJ: Princeton University Press).

Hacking, I. (1999) *The Social Construction of What?* (Cambridge, MA: Harvard University Press).

Hall, L.A. (2003) 'It was affecting the medical profession': The history of masturbatory insanity revisited. *Paedogogica Historica*, 39, 685–99.

Hare, E.H. (1962) Masturbatory insanity: The history of an idea. *Journal of Mental Science*, 108, 1–25.

Hare, E. (1987) Creativity and mental illness. *British Medical Journal (Clinical Research Edition)*, 295, 1587–9.

Harley, H. (1921) Segregation vs. hanging. *Journal of the American Institute of Criminal Law and Criminology*, 11, 512–27.

Harriman, P.L. (1947) *Dictionary of Psychology* (New York: The Philosophical Library).

Harrow, M. and Silverstein, M. (1977) Psychotic symptoms in schizophrenia after the acute phase. *Schizophrenia Bulletin*, 3, 608–16.

Harrow, M. and Quinlan, D.M. (1985) *Disordered Thinking and Schizophrenic Psychopathology* (London: Gardner Press).

Hasleton, S. (1974) A radical notion from Australia. *Schizophrenia Bulletin*, 1, 4.

Hays, P. (1978) Taxonomic map of the schizophrenias, with special reference to puerperal psychosis. *British Medical Journal*, 2, 755–7.

Hecker, E. (1871) Zur begründung des klinischen standpunktes in der psychiatrie. *Archiv für pathologische Anatomie und Physiologie und für Klinische Medicin*, 52, 203–18.

Hecker, E. (1877) Zur klinischen diagnostik und prognostik der psychischen krankheiten. *Allgemeine Zeitschrift für Psychiatrie*, 33, 602–20.

Heinze, A.R. (2003) Schizophrenia Americana: Aliens, alienists, and the "personality shift" of twentieth-century culture. *American Quarterly*, 55, 227–56.

Hell, D., Scharfetter, C. and Möller, A. (eds) (2001) *Eugen Bleuler: Leben und Werk* [*Eugen Bleuler: Life and Work*] (Bern: Hans Huber).

Hemphill, R.E. (1952) Incidence and nature of puerperal psychiatric illness. *The British Medical Journal*, 2, 1232–5.

Herbert, W. (1983) Soviet psychiatrists make preemptive bid, quit world body. *Science News*, 123, 116.

Heuyer, G. (1974) *La Schizophrénie* (Paris: Presses Universitaires de France).

Higdon, H. (1999) *Leopold & Loeb: The Crime of the Century* (Champaign, IL: University of Illinois Press).

Hilgard, E.R. (1962) *Introduction to Psychology* (New York: Harcourt, Brace & World).

Hill, G.H. (1900) Dementia praecox. *Proceedings of the American Medico-Psychological Association*, 7, 282–90.

Hill, L.B. (1955). *Psychotherapeutic Intervention in Schizophrenia* (Chicago, IL: The University of Chicago Press).

Hingtgen, J.N. and Bryson, C.Q. (1972) Recent developments in the study of early childhood psychoses; infantile autism, childhood schizophrenia, and related disorders. *Schizophrenia Bulletin*, 1, 8–54.

Hinsie, L.E. and Shatzky, J. (1940) *Psychiatric Dictionary* (London: Oxford University Press).

Hinsie, L.E. and Campbell, R.J. (1970) *Psychiatric Dictionary*, 4th ed. (London: Oxford University Press).

Hoch, A. (1912) Dementia Praecox. *Psychological Bulletin*, 9, 169–72.

Hoch, P. and Polatin, P. (1949) Pseudoneurotic forms of schizophrenia. *Psychiatric Quarterly*, 23, 248–76.

Hoche, A. (1912) Die Bedeutung der Symptomenkomplexe in der Psychiatrie. *Zeitschrift für die gesamte Neurologie und Psychiatrie*, 12, 540–51.

Hoenig, J. (1995) 'Schizophrenia', in: G. Berrios and R. Porter (eds) *A History of Clinical Psychiatry: The Origins and History of Psychiatric Disorders*, pp. 336–48 (London: Athlone Press).

Hoff, H. and Arnold, O.H. (1961) 'Germany and Austria', in: L. Bellak (ed.) *Contemporary European Psychiatry*, pp. 43–145 (New York: Grove Press).

Hofling, C.K. and Leninger, M.M. (1960) *Basic Psychiatric Concepts in Nursing*, 1st ed. (Philadelphia, PA: J.B. Lippincott).
Hofmann, G. (1963) *Experimentelle grundlagen der multifaktoriellen genese der schizophrenie* (Wien: Springer Verlag).
Hogarty, G.E. (1977) Treatment and course of schizophrenia. *Schizophrenia Bulletin*, 3, 587–99.
Holden, C. (1986) Proposed new psychiatric diagnoses raise charges of gender bias. *Science New Series*, 321, 327–8.
Holland, J. and Shakhmatova-Pavlova, I.V. (1977) Concept and classification of schizophrenia in the soviet union. *Schizophrenia Bulletin*, 3, 277–87.
Holzman, P. (1977) The modesty of nature: A social perspective on schizophrenia. *Social Service Review*, 51, 588–603.
Holzman, P.S. and Matthysee, S. (1990) The genetics of schizophrenia: a review. *Psychological Science*, 1, 279–86.
Horwitt, M.K. (1956) Fact and artifact in the biology of schizophrenia. *Science*, 124, 429–30.
Hoskins, R.G. (1946) *The Biology of Schizophrenia* (New York: W.W. Norton).
Howells, J.G. (1991) *The Concept of Schizophrenia: Historical Perspectives* (Washington, DC: American Psychiatric Press).
Howells, J.G. and Osborn, M.L. (1984) *A Reference Companion to the History of Abnormal Psychology* (London: Greenwood Press).
Hughes, E.M. (1925) Social significance of Huntington's chorea. *American Journal of Psychiatry*, 81, 537–74.
Hurst, L. (1960) Phenothiazine derivatives in the treatment of schizophrenia. *Journal of Mental Science*, 106, 755–70.
Hutchings, R.H. (1945) *A Psychiatric Word Book*, 7th ed. (Utica, NY: The State Hospitals Press).
Ion, R.M. and Beer, M.D. (2002) The British reaction to dementia praecox. 1893–1913: Part 1. *History of Psychiatry*, 13, 285–304.
Ingleby, D. (ed.) (1981). *Critical Psychiatry: The Politics of Mental Health* (New York: Penguin Books).
Ireland, W. (1902) Differential diagnosis between hysteria and katatony [*Beiträge zu differential diagnose der hysterie und katatonie*] Kaiser, O. *Journal of Mental Science [British Journal of Psychiatry]*, 48, 582.
James, G.W.B., Freudenberg, R. and Cannon, A.T. (1937) Insulin shock treatment of schizophrenia. *The Lancet*, 229, 1101–4.
Jameson, R. (1985) Personal view. *British Medical Journal*, 291, 541.
Jansson, L.B. and Parnas, J. (2006) Competing definitions of schizophrenia: What can be learned from polydiagnositic studies. *Schizophrenia Bulletin*, 33, 1178–200.
Jansson, L., Handest, P., Nielsen, J., Saebye, D. and Parnas, J. (2002) Exploring boundaries of schizophrenia: A comparison of ICD-10 with other diagnostic systems in first-admitted patients. *World Psychiatry*, 1, 109–14.
Jaspers, K. (1949/1977) *Strindberg and Van Gogh* (Tuscon, AZ: The University of Arizona Press) (Original work published 1922).
Jaspers, K. (1962). *General Psychopathology 7th edition* (J. Hoenig and M. Hamilton. Trans.).(Manchester: Manchester University Press).
Jelliffe, S.E. (1898) Zur katatonie-frage. Eine klinische studie. V. Schüle. *Journal of Nervous and Mental Disease*, 25, 632.

Jelliffe, S.E. (1927) The mental pictures in schizophrenia and in epidemic encephalitis. *American Journal of Psychiatry*, 6, 413–65.

Jelliffe, S.E. and Latimer, C.W. (1915) *Appleton's Medical Dictionary* (New York: D. Appleton).

Jelliffe, S.E. and White, W.A. (1919) *Diseases of the Nervous System; A Text-book of Neurology and Psychiatry*, 3rd ed. (Philadelphia, PA: Lea & Febiger).

Jonas, A. and Jonas, D. (1975) An evolutionary context for schizophrenia. *Schizophrenia Bulletin*, 1, 33–41.

Jörger, J.B. (1918) Über Dienstverweigerer und Friedensapostel [On Conscientious Objectors and Apostles of Peace]. *Zeitschrift für die gesamte Neurologie und Psychiatrie*, 43, 117–33.

Jung, C. (1907/1972) *Über die psychologie der dementia praecox* [*On the Psychology of Dementia Praecox*] (Glarus: Walter).

Jung, C. (1911/1972) 'A criticism of Bleuler's theory of schizophrenic negativism', in: McGuire, W. (ed.) *The Collected Works of C.G. Jung*, Vol. 3, pp. 197–203 (London: Routledge & Keegan Paul).

Jung, C. (1919/1972) 'On the problem of psychogenesis in mental disease', in: McGuire, W. (ed.) *The Collected Works of C.G. Jung*, Vol. 3, pp. 211–25 (London: Routledge & Keegan Paul).

Jung, C. (1928/1972) 'Mental disease and the psyche', in: McGuire, W. (ed.) *The Collected Works of C.G. Jung*, Vol. 3, pp. 226–33 (London: Routledge & Keegan Paul).

Jung, C. (1939/1972) 'On the psychogenesis of schizophrenia', in: McGuire, W. (ed.) *The Collected Works of C.G. Jung*, Vol. 3, pp. 233–49 (London: Routledge & Keegan Paul).

Jung, C. (1974) 'Letter to Sigmund Freud. April 17th 1910', in: McGuire, W. (ed.) *The Freud/Jung Letters: The Correspondence Between Sigmund Freud and C.G. Jung*, p. 35 (Princeton, NJ: Princeton University Press).

Kahlbaum, K.L. (1874) *Die katatonie oder das spannungsirresein* (Berlin: Hirschwald).

Kahlbaum, K.L. (1874/1973) *Catatonia* (Baltimore, MD: John Hopkins University Press).

Kalman, J. (1977) On combat neurosis: Psychiatric experience during the recent Middle-East war. *International Journal of Social Psychiatry*, 23, 195–203.

Kallman, F.J. (1938) *The Genetics of Schizophrenia* (New York: J.J. Augustin).

Kanner, L. (1943) Autistic disturbances of affective contact. *The Nervous Child*, 2, 217–50.

Karnosh, L. (1932) *A Psychiatrist's Anthology* (Cleveland, OH: The Occupational Therapy Press.)

Kasanin, J. (1933) The acute schizoaffective psychoses. *American Journal of Psychiatry*, 90, 97–127.

Kasanin, J. and Kaufmann, M.R. (1929) A study of the functional psychoses in childhood. *American Journal of Psychiatry*, 86, 307–83.

Katz, M.M., Cole, J.O. and Lowery, H.A. (1969) Studies of the diagnostic process. The influence of symptom perception, past experience, and ethnic background on diagnostic decisions. *American Journal of Psychiatry*, 125, 937–47.

Kaufman, I., Frank, T., Heims, L., Herrick, J., Reiser, D. and Willer, L. (1960) Treatment implications of a new classification of parents of schizophrenic children. *American Journal of Psychiatry*, 116, 920–4.

Kayton, L. (1972) The relationshiop of the vampire legend to schizophrenia. *Journal of Youth Adolescence*, 1, 303–14.

Keith, S.J., Gunderson, J., Reifman, A., Buchsbaum, S. and Mosher, L.R. (1976) Special report: schizophrenia 1976. *Schizophrenia Bulletin*, 2, 509–65.

Kellam, A. (1989) French empirical criteria for the diagnosis of non affective non-organic psychoses. Comparison between the criteria suggested by professors Pull and Pichot and those of DSM-III-R. *British Journal of Psychiatry*, 155, 153–9.

Kempf, E.J. (1920) *Psychopathology* (St Louis, MO: Mosby).

Kendell, R. (1972/1975) 'Schizophrenia: The remedy for diagnostic confusion', in: Silverstone, T. and Barraclough, B. (eds) *Contemporary Psychiatry: Selected Reviews from the British Journal of Hospital Medicine*, pp. 11–17 (Kent: Headley Brothers).

Kessler, S. (1976) Progress and regress in the research on the genetics of schizophrenia. *Schizophrenia Bulletin*, 2, 434–9.

Kety, S. (1959) Biochemical theories of schizophrenia. *Science*, 129, 1528–32.

Kety, S. (1974) From rationalization to reason. *American Journal of Psychiatry* 131, 957–63.

Keyes, J. and Hofling, C. K. (1974) *Basic Psychiatric Concepts in Nursing*, 3rd ed. (Philadelphia, PA: J.B. Lippincott).

Kirby, G.H. (1912/1915) Dementia praecox deteriorations without trends. *Cornell University Medical Bulletin*, 5, 1–12.

Kleist, K. and Dreist, K. (1937) Die Katatonie auf Grund katamnestischer Untersuchungen. *Zeitschrift für die Gesamte Neurologie Und Psychiatrie*, 157, 479–556.

Klerman, G. (1977) Better but not well: Social and ethical issues in the deinstitutionalization of the mentally ill. *Schizophrenia Bulletin*, 3, 617–31.

Klerman, G.L. (1978) *'The evolution of a scientific nosology'*, in: Shershow, J.C. (ed.) *Schizophrenia: Science and Practice*, pp. 99–122 (Cambridge, MA: Harvard University Press).

Klerman, G. (1989) Psychiatric diagnostic categories: issues of validity and measurement: An invited comment on Mirowsky and Ross (1989). *Journal of Health and Social Behavior*, 30, 26–32.

Kline, N. (1952) *Synopsis of Eugen Bleuler's Dementia Praecox or the Group of Schizophrenias* (New York: International Universities Press).

Klüver, H. (1926) Mescal visions and eidetic vision. *The American Journal of Psychology*, 37, 502–15.

Kohn, M. (1973) Social class and schizophrenia: A critical review and a reformulation. *Schizophrenia Bulletin*, 1, 60–79.

Konrád, E. (1885) Zur Lehre von der acuten hallucinatorischen Verworrenheit. *Archiv für Psychiatrie und Nervenkrankheiten*, 16, 522–40.

Kohnstamm, O. (1914) Schizothymie und Zyklothymie. *Berliner Klinische Wochenschrift*, 51, 1680.

Kopeloff, L.M., Kopeloff, N., Hinsie, L.E. and Etchells J.L. (1935) Tuberculous bacillemia and dementia praecox. *The Journal of Bacteriology*, 30, 49–59.

Kormos, H.R. (1978) 'The nature of combat stress', in: Figely, C.R. (ed.) *Stress Disorders Among Vietnam Veterans*, p. 4 (Kentucky, LA: Brunner and Mazel).

Kraam, A. and Phillips, P. (2012) Hebephrenia: A conceptual history. *History of Psychiatry*, 23, 387–403.

Kraepelin, E. (1887) *Psychiatrie: Ein kurzes lehrbuch für studirende und aerzte*, 2nd ed. (Leipzig: Ambr. Abel).

Kraepelin, E. (1893) *Psychiatrie: Ein kurzes lehrbuch für studirende und aerzte* [*A Short Text Book for Students and Doctors*], 4th ed. (Leipzig: Arthur Meiner).

Kraepelin, E. (1896) *Psychiatrie: Ein lehrbuch für studirende und aerzte* [*A Text Book for Students and Doctors*], 5th ed. (Leipzig: Barth).

Kraepelin, E. (1903) *Psychiatrie: Ein lehrbuch für studirende und aerzte* [*A Short Text Book for Students and Doctors*], 7th ed. (Leipzig: Barth).

Kraepelin, E. (1904/2002) *Lectures on Clinical Psychiatry* (T. Johnstone, trans.) (Bristol: Thoemmes Press).

Kraepelin, E. (1913/1919) *Dementia Praecox and Paraphrenia.* (M. Barclay. trans.) (Chicago, IL: Chicago Medical Book Company).

Kraepelin, E. (1920/1992) Der erscheinungsformen des irreseins [*The Manifestations of Insanity*]. *History of Psychiatry*, 3, 503–29.

Kramer, M. (1961) 'Some problems for international research suggested by observations on differences in first admission rates to the mental hospitals of England and Wales and of the United States', in: Rinkle, M. (ed.) *Proceedings of the Third World Congress of Psychiatry*, Vol. 3, pp. 153–60 (Montreal: University of Toronto Press).

Kretschmer, E. (1921) *Körperbau und Charakter* [*Physique and Character*] (Berlin: Julius Springer).

Kretschmer, E. (1925/1999) *Physique and Character. An Investigation of the Nature of Constitution and of the Theory of Temperament* (London: Routledge).

Kyziridis, T. (2005) Notes on the history of schizophrenia. *German Journal of Psychiatry*, 8, 42–8.

Lacan, J. (1975) *De la psychose paranoïaque dans ses rapports avec la personnalité* (Paris: Éditions du Seuil).

Ladell, R.M. (1951) *A Dictionary of Psychological Terms* (London: W. M. Samson).

Laforgue, R. (1927) Schizophrenia, Schizomania and Schizonoïa. *Revue Française de Psychanalyse*, 1, 6–20.

Laing, R.D. (1960/1990) *The Divided Self* (London: Penguin).

Laing, R.D. (1967a) 'The study of family and social contexts in relation to the origin of schizophrenia', in: Romano, J. (ed.) *The Origins of Schizophrenia*, pp. 139–46 (Amsterdam: Excerpta Medica Foundation).

Laing, R.D. (1967b) *The Politics of Experience* (New York: Pantheon Books).

Laing, R.D. and Esterson, A. (1964/1970) *Sanity Madness and the Family: Families of Schizophrenics* (London: Penguin).

Larkin, E. and Gillies, H. (1938) Prognosis of schizophrenia. *The Lancet*, 231, 461–2.

Larousse (1976). Schizophrénie. *La grande encyclopédie* (Paris: Larousse).

Langfeldt, G. (1937) *The Prognosis in Schizophrenia and the Factors Influencing the Course of the Disease* (Copenhagen: Munksgaard).

Lavretsky, H. (1998) The Russian concept of schizophrenia: A review of the literature. *Schizophrenia Bulletin*, 24, 537–57.

Lehmann, H.E. (1971) 'The impact of the therapeutic revolution on noslogy', in: Cancro, R. (ed.) *The Schizophrenic Syndrome*, pp. 136–54 (London: Butterworths).

Lehmann, H. (1975) Psychopharmacologicial treatment of schizophrenia. *Schizophrenia Bulletin*, 1, 27–45.

Lehmann, H.E. (1980a) 'Schizophrenia: Clinical features', in: Kaplan, H., Freedman, A.M. and Sadock, B.J. (eds) *Comprehensive Textbook of Psychiatry*, 3rd ed., Vol. 2, pp. 1153–92 (London: Wilkins).

Lehmann, H.E. (1980b) 'Schizophrenia history', in: Kaplan, H., Freedman, A. and Sadock, B. (eds) *Comprehensive Textbook of Psychiatry*, 3rd ed., Vol. 2, pp. 1105–13 (London: William and Wilkins).

Leppmann, A. (1890) *Die sachverständigen-thätigkeit bei seelenstörungen: ein kurzgefasstes handbuch für die ärztliche Praxis* (Berlin: R. Schoetz).

Leuba, J. (1936) La famille névrotique et les névrose familiales. *Revue Française de Psychanalyse*, 9, 360–419.

Leudar, I. and Thomas, P. (2000) *Voices of Reason, Voices of Insanity* (Hove: Bruner-Routledge).

Levin, M. (1946) Transitory schizophrenias produced by bromide intoxication. *American Journal of Psychiatry*, 103, 229–37.

Lewin, B. (1934) Towards mental health: The schizophrenic problem. *Psychoanalytic Quaterly*, 3, 317–25.

Lewis, A. (1979) *The Later Papers of Aubrey Lewis* (Oxford: Oxford University Press).

Lewis, N.D.C. (1925/1928) 'Graphic art productions', in: Dana, C.L., Davis, T.K., Jelliffe, S.E., Riley, H.A., Tilney, F. and Timme, W. (eds) *Schizophrenia [Dementia Praecox]; An Investigation by the Association for Research in Nervous and Mental Diseases*, pp. 344–68 (New York: Paul B. Hoeber).

Lewis, N.D.C. (1936) *Research in Dementia Praecox* (New York: National Committee for Mental Hygiene).

Lewis, W.B. (1889) *A Textbook of Mental Diseases. The Pathological Aspects of Insanity* (London: Charles Griffin).

Lewis, N.D.C. and Blanchard, E. (1931) Clinical findings in 'recovered' cases of schizophrenia. *American Journal of Psychiatry*, 88, 481–92.

Lewis, A. and Minski, L. (1935) Chorea and psychosis. *The Lancet*, 225, 536–8.

Leys, R. (2000) *Trauma: A Genealogy* (Chicago, IL: University of Chicago Press).

Lidz, T. (1976) Commentary on 'a critical review of recent adoption, twin, and family studies of schizophrenia: behavioural genetics perspectives'. *Schizophrenia Bulletin*, 2, 402–12.

Lidz, T. (1978) 'A developmental theory', in: Shershow, J. (ed.) *Schizophrenia: Science and Practice*, p. 70 (Cambridge, MA: Harvard University Press).

Livet, M. (1914) Syndrome de ganser chez un détenu militaire. *Revue de Psychiatrie et de Psychologie Expérimentale*, 18, 197–208.

Lombardo, G.P. and Foschi, R. (2008) Escape from the dark forest: The experimentalist standpoint of Sante De Sanctis' psychology of dreams. *History of the Human Sciences*, 21, 45–69.

Lorr, M., Klett, C.J. and McNair, D.M. (1963) *Syndromes of Psychosis* (New York: The Macmillan Co.).

Lundbeck, E. (1994) *The Psychiatric Persuasion: Knowledge, Gender and Power in Modern America* (Princeton, NJ: Princeton University Press).

Lundholm, H. (1932) *Schizophrenia* (Durham, NC: Duke University Press).

Lurje, W. (1921) Autismus und Buddhismus. Eine parallele. *Zeitschrift für die gesamte Neurologie und Psychiatrie*, 70, 25–35.

McCulloch, J. (1995) *Colonial Psychiatry and 'The African Mind'* (Cambridge: Cambridge University Press).

McGlashan, T.H. (1988). A selective review of recent North American long-term follow-up studies of schizophrenia. *Schizophrenia Bulletin*, 14, 515–42.

McGlashan, T.H. and Fenton, W.S. (1994) 'Classical subtypes for schizophrenia', in: Widiger, T.A., Frances, A.J., Pincus, H.A., First, M.B., Ross, R. and Davis, W. (eds) *DSM-IV Sourcebook*, Vol. 1, pp. 419–41 (Washington, DC: American Psychiatric Association).

MacMahon, B. (1961) 'From the point of view of epidemiology', in: Zubin, J. (ed.) *Field Studies in the Mental Disorders*, pp. 324–32 (New York: Grune & Stratton).

McNally, K. (2007) Schizophrenia as split personality/Jekyll and Hyde: The origins of the informal usage in the English language. *Journal of the History of the Behavioral Sciences*, 43, 69–79.

McNally, K. (2009) *A History of the Concept of Schizophrenia* (PhD Dissertation), University College of Dublin.

Maier, H. (1922) Über versicherungshebephrenien. *Zeitschrift für die gesamte Neurologie und Psychiatrie*, 78, 442–53.

Martin, J.B. (1959) *The Pane of Glass* (New York: Harpers and Brothers).

Masson, M. and Azorin, J. (2007) The French mentally ill in World War II. *International Journal of Mental Health*, 35, 26–39.

Matza, D. (1968) Reviewed work: Being mentally ill by Thomas J. Scheff. *Journal of Health and Social Behavior*, 9, 346–8.

May, J.V. (1931) The dementia praecox-schizophrenia problem. *American Journal of Psychiatry*, 88, 401–46.

Mayer-Gross, W., Slater, E. and Roth, M. (1955) *Clinical Psychiatry* (Baltimore, MD: Williams & Wilkins).

Mayer-Gross, W., Slater, E. and Roth, M. (1960) *Clinical Psychiatry*, 2nd ed. (London: Cassell & Company).

Meehl, P.E. (1973) *Psychodiagnosis. Selected Papers* (Minnesota, MN: University of Minnesota Press).

Meehl, P.E. (1986) 'Diagnostic taxa as open concepts: Metatheoretical and statistical questions about reliability and construct validity in the grand strategy of nosological revision', in: Millon, T. and Klerman, G.L. (eds) *Contemporary Directions in Psychopatholgy. Towards the DSM-IV*, pp. 215–33 (New York: The Guilford Press).

Mendel, W.M. (1976) *Schizophrenia. The Experience and its Treatment* (San Francisco: Jossey-Bass).

Menninger, K. (1922) Reversible schizophrenia. A study of the implications of delirium schizophrenoides and other post influenzal syndromes. *American Journal of Psychiatry*, 78, 573–88.

Meyer, A. (1896) Psychiatrie: Ein Lehrbuch für Studirende und Aerzte. Von Dr. Emil Kraepelin [Psychiatry a textbook for students and doctors by Emil Kraepelin]. *American Journal of Insanity*, 53, 298–302.

Meyer, A. (1908) The problems of mental reaction-type, mental causes and diseases. *Psychological Bulletin*, 5, 245–61.

Meyer, A. (1910) The nature and conception of dementia praecox. *Journal of Abnormal Psychology*, 5, 247–85.

Meyer, A. (1925/1928) 'The evolution of the dementia praecox concept', in: Dana, C.L., Davis, T.K., Jelliffe, S.E., Riley, H.A., Tilney, F. and Timme, W. (eds)

Schizophrenia [Dementia Praecox]; An Investigation by the Association for Research in Nervous and Mental Diseases, pp. 3–16 (New York: Paul B. Hoeber).
Meyer, A. (1957) *Psychobiology: A Science of Man* (E. Winters, trans.) (Springfield, IL: Thomas).
Meyers, M.K. (1924) *Lang's German–English Dictionary of Terms Used in Medicine and the Allied Sciences*, 3rd ed. (London: J. & A. Churchill).
Micale, M. (1995) *Approaching Hysteria. Disease and its Interpretations* (Princeton, NJ: Princeton University Press).
Miles, B. (2002) *Jack Kerouac King of the Beats: A Portrait* (London: Virgin books).
Milici, P. (1939) Postemotive schizophrenia. *Psychiatric Quarterly*, 13, 278–93.
Milici, P. and von Salzen, C. (1938) Situational schizophrenia. *Psychiatric Quarterly*, 12, 650–68.
Minkowski, E. (1927) *La schizophrénie* [*Schizophrenia*] (Paris: Payot).
Mira, E. (1943) *Psychiatry in War* (New York: W. W. Norton).
Misra, P.C. and Hay, G.G. (1971) Encephalitis presenting as acute schizophrenia. *British Medical Journal*, 1, 532–3.
Moffson, A. (1955) A study of 400 consecutive male Bantu admissions to Weskoppies Mental Hospital. *South African Medical Journal*, 29, 689–92.
Money, J. (1974) 'Intersexual and transsexual behaviour and syndromes', in: Arieti, S. (ed.) *American Handbook of Psychiatry*, 2nd ed., Vol. 3, pp. 334–51 (New York: Basic Books).
Montandon, G. (1941) Comment reconnaître les Juifs. *Le Matin*, Aug 5, 1–2.
Morgan, J.B. (1928) *The Psychology of Abnormal People: With Educational Applications* (London: Longmans, Green & Co.).
Morriarty, J. and Masset, L. (1970) Prenatal and birth complications linked by schizophrenia research. *Science News*, 98, 15–16.
Mosher, L. (1969) The center for studies of schizophrenia. *Schizophrenia Bulletin*, 1, 4–7.
Mosher, L. (1999) Soteria and other alternatives to acute psychiatric hospitalization. A personal and professional review. *The Journal of Mental Disease*, 187, 142–9.
Moss, F.A. and Hunt, T. (1932) *Foundations of Abnormal Psychology* (New York: Prentice Hall).
Mott, F.W. and Such, M.P. (1922) Further pathologial studies in dementia praecox, especially in relation to the interstitial cells of Leydig. *Dementia Praecox Studies*, 5, 47–65.
Murphy, H. (1977) Book review: Gilles De La Tourette's syndrome: Cross-cultural analysis and treatment outcome by F.S. Abuzzahab and F.O. Andreson. *Transcultural Psychiatry*, 14, 42–3.
Murray, H. (2014) My place was set at a terrible feast. The meanings of the 'anti-psychiatry' movement and responses in the United States, 1970s–1990s. *The Journal of American Culture*, 37, 37–51.
Murray, R. and Lewis, S.W. (1987) Is schizophrenia a neurodevelopmental disorder. *British Medical Journal (Clinical Research Edition)*, 19, 295, 61–2.
Naumburg, M. (1950) *Schizophrenic Art. Its Meaning in Psychotherapy* (New York: Grune & Stratton).
Neumann, H. (1859) *Lehrbuch der psychiatrie* (Erlangen: Enke).
Noll, R. (1983) Shamanism and schizophrenia: A state-specific approach to the 'schizophrenia metaphor' of shamanic states. *American Ethnologist*, 10, 443–59.

Noll, R. (2007) Kraepelin's 'lost biological psychiatry'? Autointoxication, organotherapy and surgery for dementia praecox. *History of Psychiatry*, 18, 301–19.

Noll, R. (2011) *American Madness: The Rise and Fall of Dementia Praecox* (Boston, MA: Harvard University Press).

Noll, R. (2015) 'Centers and peripheries of liminal concepts in a profession unconcerned with diagnostic error: American alienism, prodromes and psychosis, 1894 to 1934', paper presented at 'Psychopathological Fringes: Historical and Social Science Perspectives on Category Work in Psychiatry', Institute für Geschichte der Medizin und Ethik in der Medizin, Charité-Universitätsmedizin, Feb 13, Berlin, Germany.

Neisser, C. (1887) *Über die katatonie. Ein beitrag zur klinischen psychiatrie* (Stuttgart: Von Ferdinand Enke).

Nicol, S.E. and Gottesman, I. (1983) Clues to the genetics and neurobiology of schizophenia. *American Scientist*, 71, 398–404.

Nisot, M. (1929) *La question eugénique dans les divers pays, tome II* (Brussels: Van Campenhout).

Norman, C. (1904) Dementia praecox. *British Medical Journal*, 2, 972–6.

Nuffield, E. J. (1954) The schizogenic mother. *Medical Journal of Australia*, 2, 283–6.

O'Brien, P. (1978) *The Disordered Mind. What We Know About Schizophrenia* (Upper Saddle River, NJ: Prentice Hall).

Odier, C. (1939) Doctoresse Minkowska, *Revue Française de Psychanalyse*, 11, 156–419.

Oltmanns, T.F., Neale, J.M. and Davison, G.C. (eds) (1986) *Case Studies in Abnormal Psychology*, 2nd ed. (New York: John Wiley).

Opler, M. (1959) Book reviews. *International Journal of Social Psychiatry*, 5, 236–8.

Ortega, M. (1974) Chronic Schizophrenia-readers' opinions. *Schizophrenia Bulletin*, 1, 5–6.

Osborne, R.J. (1940) A re-evaluation of the concept of schizophrenia. *Journal of Mental Science*, 86, 1078–85.

Osmond, H. and Smythies, J. (1951) Mescaline and schizophrenia. *British Medical Journal*, 2, 607.

Ougrin, D., Gluzman, S. and Dratcu, L. (2006) Psychiatry in post-communist Ukraine: Dismantling the past, paving the way for the future. *Psychiatric Bulletin*, 30, 456–9.

Overall, J.E. (1979) Brief psychiatric rating scale. *Current Contents/Social and Behavioural Sciences*, 2, 10.

Overall, J.E. and Gorham, D.R. (1962) The brief psychiatric rating scale. *Psychological Report*, 10, 799–812.

Patterson, R. (1940) The pre-psychotic personality. *Review of Educational Research*, 10, 460–3.

Peterson, F. (1897) Katatonia (katatonie of Kahlbaum – katatonische verrucktheit of Schüle *Medical Record*, 473.

Peterson, F. and Langdon, C. (1898) 'Katatonia: katatonie of Kahlbaum katatonische verucktheit of Schüle', in: Proceedings of American Medico-Psychological Association, 53rd Annual Meeting, Baltimore, May 11–14, 1897. Available at: http://www21.us.archive.org/stream/proceedingsamer15assogoog/proceedingsamer15assogoog_djvu.txt (accessed 2 January 2008)

Philsbury, W.B. (1932) *An Elementary Psychology of the Abnormal* (New York: McGraw-Hill).

Pichon, E. (1936) Freud à Dalbiez. *Revue Française de Psychanalyse*, 9, 559–88.
Pichot, P.P. (1967) Recent developments in French psychiatry. *British Journal of Psychiatry*, 113, 11–18.
Pichot, P., Guelfi, J.D. and Kroll, J. (1983) 'French perspectives on DSM- II', in: Spitzer, R.L., Williams, J.B. and Skodol, A.E. (eds) *International Perspectives on DSM III*, pp. 235–42 (Washington, DC: American Psychiatric Press).
Piquemal, J. (1912) *Considérations générales sur le réflexe. De la réflectivité tendineuse et cutanée dans la démence précoce* (Toulouse, Dirion).
Plomin, R. (1990) The role of inheritance in behaviour. *Science, New Series*, 248, 183–8.
Pollock, H. (1925/1928) 'Frequency of schizophrenia in relation of sex, age, environment, nativity and race', in: Dana, C.L., Davis, T.K., Jelliffe, S.E., Riley, H.A., Tilney, F. and Timme, W. (eds) *Schizophrenia [dementia Praecox] An Investigation by the Association for Research in Nervous and Mental Diseases*, pp. 47–69 (New York: Paul B. Hoeber).
Porot, M., Couadau, A. and Aubin, B. (1968) 'Aspects évolutifs actuels des schizophrénies', in: *Confrontations psychiatriques: Évolution de la schizophrénie*, pp. 53–77 (Paris: Specia).
Porter, R. and Micale, M. (1994) 'Reflections on psychiatry and its histories', in: Micale, M. and Porter, R. (eds) *Discovering the History of Psychiatry*, pp. 3–36 (Oxford: Oxford University Press).
Post, F. (1963) Schizophrenia. *British Medical Journal*, 1, 1734.
Postel, J. and Allen, D.F. (1994) 'History and anti-psychiatry in France', in: Micalem M. and Porter, R. (eds) *Discovering the History of Psychiatry*, pp. 384–415 (Oxford: Oxford University Press).
Price, F.W. (ed.) (1934) *A Textbook of the Practice of Medicine*, 4th ed. (London: Oxford Medical Publications).
Prince, M. (1906) *The Dissociation of a Personality: A Biographical Study in Abnormal Psychology* (New York: Longmans Green).
Prinzhor, H. (1922/1971) *Artistry of the Mentally Ill* (Berlin: Springer-Verlag).
Pull, C. and Chaillet, G. (1994) 'The nosological views of French-speaking psychiatry', in: Mezzich, J.E., Honda, Y. and Kastrup, M. (eds) *Psychiatric Diagnosis: A World Perspective*, pp. 22–32 (New York: Springer).
Putnam, F.W. (1989) *Diagnosis and Treatment of Multiple Personality Disorder* (London: The Guilford Press).
Quastel, J.H. and Wales, W.T. (1938) Faulty detoxification in schizophrenia: Abnormal excretion of hippuric acid after administration of benzoate. *The Lancet*, 232, 301–5.
Rabkin, J. (1974) Public attitudes toward mental illness: a review of the literature. *Schizophrenia Bulletin*, 1, 9–33.
Racamier, P. (1957) 'Application de la psychanalyse aux schizophrenes', in: Müller, C. and Benedetti, G. (eds) *Symposium International sur la Psychothérapie de la Schizophrénie*, pp. 129–47 (Basel: Karger).
Rapaport, D. (1945) *Diagnostic Psychological Testing. The Theory, Statistical Evaluation, and Diagnostic Application of a Battery of Tests* (Chicago, IL: The Year Book Publishers).
Rapaport, D. (1951) *Organization and Pathology of Thought* (New York: Columbia University Press).

Redlich, F. (1952) 'The concept of schizophrenia', in: Brody, E. and Redlich, F. (eds) *Psychotherapy with Schizophrenics*, p. 22 (New York: International Universities Press).

Régis, E. (1914) *Précis de psychiatrie*, 5th ed. (Paris: Doin).

Reiss, D. (1974) Competing hypotheses and warring factions: applying knowledge of schizophrenia. *Schizophrenia Bulletin*, 1, 7–11.

Reiss, D. (1975) Families and the etiology of schizophrenia. Fishing without a net. *Schizophrenia Bulletin*, 1, 9–11.

Reiss, D. and Wyatt, J. (1975) Families and biological variables in the same etiologic studies of schizophrenia: A proposal. *Schizophrenia Bulletin*, 1, 64–81.

Richards, A. (1999) A.A. Brill and the politics of exclusion. *Journal of the American Psychoanalytic Association*, 47, 9–28.

Richards, G. (1989) *On Psychological Language and the Physiomorphic Basis of Human Nature* (London: Routledge).

Richards, G. (1992) *Mental Machinery. Part 1. The Origins and Consequences of Psychological Ideas from 1600 to 1850* (London: The Athlone Press).

Richards, G. (2002) *Putting Psychology in its Place: A Critical Historical Introduction*, 2nd ed. (London: Routledge).

Robin, G. (1929) L'onanisme chez l'enfant. *Evolution psychiatrique*, 2, 87–120.

Robinson, A. (1960) Communicating with schizophrenic patients. *The American Journal of Nursing*, 60, 1120–3.

Robinson, R. (1950) *Definition* (Oxford: Clarendon Press).

Röhricht, F. and Priebe, S. (2002) Do cenesthesias and body image aberration characterize a subgroup in schizophrenia? *Acta Psychiatrica Scandinavica*, 105, 276–82.

Rojas, D.R., Outes, M. and Cositore, A. (1996) Sante de Sanctis y la demencia precocísima [Santa de Sanctis and dementia precocísima]. ALCMEON. Available at: http://www.alcmeon.com.ar/5/19/index.htm (accessed 2 January 2008).

Rollin, H.R. (1979) Schizophrenia. *British Medical Journal*, 1, 1773–5.

Romano, J. (1977) On the nature of schizophrenia: Changes in the observer as well as the observed (1932–77). *Schizophrenia Bulletin*, 3, 532–59.

Romasenko, V.A. (1953) [Histopathological modifications in hypertoxic schizophrenia]. *Zh Nevropatol Psikhiatr Im S S Korsakova*, 53, 762–9 (article in undetermined language).

Root, H.K. (1865) *The People's Medical Lighthouse* (New York: Adolphus Ranney).

Rose, N. (1999) *Governing the Soul: The Shaping of the Private Self*, 2nd ed. (London: Free Association Books).

Rosenhan, D.L. (1973) On being sane in insane places. *Science*, 179, 250–8.

Rosenthal, D. (ed.) (1963) *The Genain Quadruplets* (New York: Basic Books).

Rosenthal, D. (1969) Problems of sampling and diagnosis in the major twin studies of schizophrenia. *Schizophrenia Bulletin*, 1, 11–25.

Rosenthal, D. (1972) Schizophrenia. *Science News*, 102, 114.

Rosenthal, D. (1977) Searches for the mode of genetic transmission in schizophrenia: reflections and loose ends. *Schizophrenia Bulletin*, 3, 268–76.

Ross, D. (1972) *G. Stanley Hall. The Psychologist as Prophet* (Chicago, IL: The University of Chicago press).

Roth, R. (1977) Psychiatrists: are they pagan priests of a sacred symbol? An interview with Dr. Thomas Szasz. *Modern Medicine*, 2, 19–20.

Roth, M. (1978) 'Problems of diagnosis and treatment in the borderlands of schizophrenia', in: Simister, J.M. (ed.) *Neuroleptics and Schizophrenia*, pp. 63–77 (Luton: Lundbeck House).

Rothblum, E.D, Solomon, L.J. and Albee, G.W. (1986) 'A socio-political perspective of DSM-III', in: Millon, T. and Klerman, G.L. (eds) *Contemporary Directions in Psychopathology. Towards the DSM-IV*, pp. 73–99 (New York: The Guilford Press).

Roudinesco, E. (1986) *Jacques Lacan & co. A History of Psychoanalysis in France, 1925–1985* (Chicago, IL: The University of Chicago Press).

Rowe, D. (1980) Philosophy and psychiatry. *Philosophy*, 55, 109–112.

Rust, G. (1878) *Uber die katatonie oder das spannungsirresein* (Berlin: no publisher given).

Ruiz, D.S. (1982) Epidemiology of schizophrenia: Some diagnostic and sociocultural considerations. *Phylon*, 43, 315–26.

Sakel, M. (1958) *Schizophrenia* (New York: Philosophical Library).

Salzinger, K. (1973) *Schizophrenia: Behavioural Aspects* (New York: John Wiley).

Sarbin, T. (1972) Schizophrenia is a myth, born of metaphor, meaningless. *Psychology Today*, 6, 18–27.

Sarbin, T.R. and Mancuso, J.C. (1980) *Schizophrenia: Medical Diagnosis or Moral Verdict?* (New York: Pergamon Press).

Sass, L. (1987) Introspection, schizophrenia, and the fragmentation of self. *Representations*, 19, 1–34.

Scharfetter, C. (1999) *Schizophrene menschen: diagnostic, psychopathologie, forschungsansatze*, 5th Auflage [*Schizophrenic People: Diagnosis, Psychopathology, Basic Research*, 5th ed.] (Weinheim: Psychologie Verlags Union).

Scharfetter, C. (2006) *Eugen Bleuler 1857–1939. Polyphrenie und Schizophrenie* [*Eugen Bleuler 1857–1939. Polyphrenia and Schizophrenia*] (Zurich: VDF Hochschulverlag).

Scheff, T. (1966) *Being Mentally Ill* (Chicago, IL: Adeline Publishing Company).

Scheff, T. (1970) Schizophrenia as ideology. *Schizophrenia Bulletin*, 1, 15–19.

Scheff, T. (1973) 'Schizophrenia as ideology', in: P. Brown (ed.) *Radical Psychology*, pp. 46–60 (New York: Evanston).

Schiff, P. (1937) La prophylaxie criminelle en dehors de la prison. *Annales de Médecine Légale, de Criminologie et de Police Scientifique*, 1, 290–348.

Schneider, K. (1918) Schizophrene Kriegpsychose. *Zeitschrift für Neurologie*, 43, 420–39.

Schneider, K. (1959) *Clinical Psychopathology* (M.W. Hamilton, trans.) (New York: Grune and Stratton).

Schüle, H. (1888) *Traité Clinique des maladies mentales* (translated from 3rd German ed. by J. Dagonet and G. Duhamel) (Paris: Lecrosnier et Babé).

Schulhof, F. (1928) *Schizo-phrenie schizo-bulie* (Leipzig: Franz Deuticke).

Schulz, C. (1975) An individualized psychotherapeutic approach with the schizophrenic patient. *Schizophrenia Bulletin*, 1, 46–69.

Seidelman, W. (1988) Mengele medicus: Medicine's Nazi heritage. *The Milbank Quarterly*, 66, 221–39.

Shamdasani, S. (2005) *Jung and the Making of Modern Psychology. The Dream of a Science* (Cambridge: Cambridge University Press).

Shapley, D. (1974) U.S.–U.S.S.R Exchange: Americans split on schizophrenia program. *Science, New Series*, 183, 932–5.

Shaw, D., Bloch, S. and Vickers, A. (1972) Psychiatry and the state. *New Scientist*, 56, 258–61.

Shean, G.G. (2004) *Understanding and Treating Schizophrenia. Contemporary Research Theory and Practice* (New York: The Haworth Press).

Shershow, J. (ed.) (1978) *Schizophrenia: Science and Practice* (Cambridge, MA: Harvard University Press).

Shorter, E. (2005) *A Historical Dictionary of Psychiatry* (Oxford: Oxford University Press).

Silverman, J. (1967) Shamans and acute schizophrenia. *American Antrhopologist*, 69, 21–31.

Simson, T.P. (1960) Infektsionnye psikhozy (infection psychoses): translated from the Russian. *Public Health Reports*, 75, 451–6.

Singh, B. (2005) Recognition and optimal management of schizophrenia and related psychoses. *Internal Medicine Journal*, 35, 413–18.

Skottowe, I. (1940) Prognosis of schizophrenia. *British Medical Journal*, 2, 613.

Skrabanek, P., Breakey, W.R. and Kala, A.K. (1977) Typhoid catatonia and ECT. *British Medical Journal*, 2, 892–3.

Slaughter, M.M. (1982) *Universal Languages and Scientific Taxonomy in the Seventeenth Century* (Cambridge: Cambridge University Press).

Smith, R.P. (1892) Cases of so-called katatonia. *Journal of Mental Science [British Journal of Psychiatry]*, 38, 403–7.

Soccaras, F (1957) Schizophrénie pscudo-névrotique et schizophrénie pseudo-caractérielle. *Revue française de psychanalyse*, 21, 535–51.

Solomon, M. (1917) Need for a stricter definition of terms in psychopathology. *The Journal of Abnormal Psychology*, 12, 195–9.

Southard, E.E. (1910) A study of the dementia praecox group in the light of certain cases showing anomalies or scleroses in particular brain-regions. *American Journal of Psychiatry*, 67, 119–76.

Southard, E.E. (1914) Cortex lesions in dementia praecox. *The American Journal of Insanity*, 71, 383–403.

Southard, E.E. and Noll, R. (2007) Classic Text No. 72. Non-dementia non-praecox: Note on the advantages to mental hygiene of extirpating a term. *History of Psychiatry*, 18, 483–502.

Spittell, A. (ed.) (1979) *Tice's Practice of Medicine*, Vol. 10 (Baltimore, MD: Harper Row).

Spitzer, R.L. (1975) On pseudoscience in science, logic in remission, and psychiatric diagnosis: a critique of Rosenhan's 'On Being Sane in Insane Places'. *Journal of Abnormal Psychology*, 84, 442–52.

Spitzer, R.L., Endicott, J. and Robins, E. (1975) *Research Diagnostic Criteria (RDC)* (New York: New York State Psychiatric Institute).

Spitzer, R.L., Andreason, N.C. and Endicott, J. (1978) Schizophrenia and other psychotic disorders. *Schizophrenia Bulletin*, 4, 489–511.

Spitzka, E.C. (1887) Cases of masturbation (Masturbatic Insanity), *Journal of Mental Science [British Journal of Psychiatry]*, 33, 57–73.

Squires, P.C. (1937) Fyodor Dostoevsky a psychopathographical sketch. *Psychoanalytic Review*, 24, 365–89.

Stauder, K.H. (1934) Die tödliche Katatonie. *Archiv für Psychiatrie und Nervenkrankheiten*, 102, 614–34.

Stedman, T.L. (1911) *A Practical Medical Dictionary* (Oxford: Oxford UniversityPress).
Steiner, C. (1915). Do other strange personalities lurk hidden in this clergyman? *Washington Post*, Feb 14, p. 4.
Stengel, E. (1957) Concepts of schizophrenia. *British Medical Journal*, 1, 1174–6.
Stephens, J.H. (1978) Long-term prognosis and followup in schizophrenia. *Schizophrenia Bulletin*, 4, 25–47.
Stevenson, R.L. (1886/1994) *The Strange Case of Dr. Jekyll and Mr. Hyde* (London: Penguin).
Stoddart, W.H. (1909) *Mind and its Disorders; A Text-book for Students and Practitioners* (Philadelphia, PA: Blakiston).
Stotz-Ingenlath, G. (2000) Epistemological aspects of Eugen Bleuler's conception of schizophrenia in 1911. *Medicine, Health Care and Philosophy*, 3, 153–9.
Stone, H.K. and Dellis, N.P. (1960) An exploratory investigation into the levels hypothesis. *Journal of Projective Techniques*, 24, 333–40.
Strauss, J.S., Carpenter, W.T. and Bartko, J. (1974a) The diagnosis and understanding of schizophrenia. Part I. Use of signs and symptoms for the identification of schizophrenicpatients. *Schizophrenia Bulletin*, 1, 37–49.
Strauss, J.S., Carpenter, W.T. and Bartko, J. (1974b) The diagnosis and understanding ofschizophrenia. Part III. Speculations on the process that underline schizophrenic symptoms and signs. *Schizophrenia Bulletin*, 1, 61–9.
Strecker, E.A. and Ebaugh, F.G. (1925/1926) Psychoses occurring during the puperium. *Archives of Neurology and Psychiatry*, 15, 239–52.
Sullivan, H.S. (1931) The modified psychoanalytic treatment of schizophrenia. *American Journal of Psychiatry*, 88, 519–40.
Sullivan, H.S. (1938/1965) *Clinical Studies in Psychiatry* (New York: Norton).
Sullivan, H.S. (1939/1953) *Conceptions of Modern Psychiatry*, 2nd ed. (New York: W.W. Norton).
Sullivan, H.S. (1962) *Schizophrenia as a Human Process*, 1st ed. (New York: W.W. Norton).
Susman, W. (1984) *Culture as History. The Transformation of American Society in the Twentieth Century* (New York: Pantheon Books).
Sutherland, S. (1976) *Breakdown. A Personal Crisis and a Medical Dilemma* (London: Weidenfeld and Nicolson).
Szasz, T. (1957) The problem of psychiatric nosology. *American Journal of Psychiatry*, 114, 405–13.
Szasz, T.S. (1960). The myth of mental illness. *American Psychologist*, 15, 113–18.
Szasz, T.S. (1976) Schizophrenia, the sacred symbol of Psychiatry. *The British Journal of Psychiatry*, 129, 308–16.
Szasz, T. (1976/1988) *Schizophrenia: The Sacred Symbol of Psychiatry* (New York: Syracuse University Press).
Szasz, T. (2008) Debunking antipsychiatry: Laing, law, and Largactil. *Current Psychology*, 27, 79–101.
Szasz, T. (2009). *Antipsychiatry: Quackery Squared* (New York: Syracuse Press).
Taylor M.A., Shorter E., Vaidya N.A., Fink M. (2010) The failure of the schizophrenia concept and the argument for its replacement by hebephrenia: Applying the medical model for disease recognition. *Acta Psychiatrix Scandinavica* 22, 173–83.
Teller, D.N. and Denber, H.C. (1968) Defining schizophrenia with the techniques of molecular biology. *Disease of the Nervous System*, 29, 93–112.

Tétry, A. (1948) L'hérédité. Buts et méthhodes de la génétique humain. Par le Dr F. Twiesselmann. *Revue scientifique*, 1, 63–4.

Thigpen, C. and Cleckley, H. (1954) A case of multiple personality. *Journal of Abnormal and Social Psychology*, 49, 135–51.

Thigpen, C. and Cleckley, H. (1957) *The Three Faces of Eve* (London: Secker and Warburg).

Time (1939) Death for sanity. *Time*, Nov 20. Available at: http://www.time.com/time/magazine/article/0,9171,762822,00.html (accessed 15 May 2008).

Time (1948) Frosted children. *Time*, Apr 26. Available at: http://www.time.com/time/magazine/article/0,9171,798484,00.html (accessed 15 May 2007).

Torrey, E.F. (1973) Is schizophrenia universal? An open question. *Schizophrenia Bulletin*, 1, 53–9.

Torrey, E.F. (1983) *Surviving Schizophrenia: A Family Manual* (New York: Harper & Row).

Torrey, E.F. (1988) Stalking the schizovirus. *Schizophrenia Bulletin*, 14, 223–9.

Toulouse, E. and Mourgue, R. (1925) Le caractère social de l'hygiène et de la prophylaxie mentales. *La Prophylaxie Mentale*, 4, 150–61.

Travis, L.E. (1924) Test for distinguishing between schizophrenoses and psychoneuroses. *Journal of Abnormal and Social Psychology*, 19, 283–98.

Trotter, R. (1972). Schizophrenia. *Science News*, 102, 58–9.

Trotter, R. (1977) Schizophrenia: A cruel chain of events. *Science News*, 111, 394–5.

US Department of Health, Education and Welfare. (1972) *ICDA: Eight International Classification of Diseases, Adapted for use in the United States* (Washington, DC: Public Health Service).

Vaillant, G.E. (1978a) Prognosis and the course of schizophrenia. *Schizophrenia Bulletin*, 4, 20–4.

Vaillant, G.E. (1978b) A 10-year followup of remitting schizophrenics. *Schizophrenia Bulletin*, 4, 78–85.

Van Praag, H. (1978) Comments on the impossible concept of schizophrenia. *Psychiatric Praxis*, 5, 73–87.

Van Teslaar, J.S. (1912) Dementia praecox oder gruppe der schizophrénie [dementia praecox or the group of schizophrenias]. *Journal of Abnormal Psychology*, 7, 374–77.

Volmat, R. (1958). *La schizophrénie par l'image* (Paris: Laboratoires Roche).

Von Bueltzingsloewen, I. (2007) *L'hécatombe des fous. La famine dans les hôpitaux psychiatriques français sous l'occupation* (Paris: Flammarion).

Vonnegut, M. (1974). Why I want to bite R. D. Laing. *Harper's Magazine*, 248, 90–2.

War Department (1940). *Outline of Neuropsychiatry in Aviation Medicine: Technical Manual 8–325* (Washington, DC: War Department).

Warnock, J. (1895) A case of catalepsy, with prolonged silence, alternating with verbigeration. *Journal of Mental Science*, 41, 82–6.

Warren, H.C. (1934) *Dictionary of Psychology* (Cambridge: The Riverside Press).

Washington Post (1916) He calls it schizophrenia and places blame on war. *Washington Post*, Jul 16, p. A5.

Washington Post (1919) Was Jekyll and Hyde. *Washington Post*, Jul 17, p. 4.

Washington Post (1924a) Leopold and Loeb at mental border line says doctor. *Washington Post*, Aug 17, p. 3.

Washington Post (1924b) Childish phantasy theory ridiculed in Franks hearing. *Washington Post,* Aug 3, p. 1.

Washington Post (1931) Scientist raps procedure in Lincoln Psychoanalysis. *Washington Post,* Jun 6, p. 2.

Weinberg, S.K. (1950) A sociological analysis of a schizophrenic type. *American Sociological Review,* 15, 600–10.

Weiner, H. (1980) 'Schizophrenia: Etiology', in: Kaplan, H., Freedman, A. and Sadock, B. (eds) *Comprehensive Textbook of Psychiatry,* 3rd ed., Vol. 3, pp. 1121–53 (London: Williams & Wilkins).

Wells, F.L. (1946) Hölderlin: Greatest of 'schizophrenics'. *The Journal of Abnormal and Social Psychology,* 41, 199–206.

Whitaker, L.C. (1992) *Schizophrenic Disorders. Sense and Nonsense in Conceptualisation, Assessment, and Treatment* (London: Plenum Press).

Wick, S. (1940) Schizophrenia. *The American Journal of Nursing,* 40, 1071–6.

Wickes, I. (1954) Foetal defects following insulin coma therapy in early pregnancy. *British Medical Journal,* 2, 1029–30.

Wilkerson, T.E. (1995) *Natural Kinds* (Aldershot: Avebury).

Will, O.A. (1980) 'Schizophrenia: Psychological treatment', in: Kaplan, H., Freedman, A. and Sadock, B. (eds) *Comprehensive Textbook of Psychiatry,* 3rd ed., Vol 2, pp. 1217–40 (London: Williams & Wilkins).

Wille, L. (1888) Die Lehre von der Verwirrtheit. *Archiv für Psychiatrie und Nervenkrankheiten,* 19, 328–51.

Wilson, H. (1951) Refresher couse for general practioners. The early diagnosis of schizophrenia. *British Medical Journal,* 1, 1502–4.

Wilson, I. (1935) London association of the medical women's federation; Feb 9th. *The Lancet,* 225, 327.

Wilson, M. (1993) DSM-III and the transformation of American psychiatry: A history. *American Journal of Psychiatry,* 150, 399–410.

Wilson, P.J. (1978) The outcast and the prisoner: Models for witchcraft and schizophenia. *Man, New Series,* 13, 88–99.

Wing, J.K. (1961) A simple and reliable subclassification of chronic schizophrenia. *The British Journal of Psychiatry,* 107, 862–75.

Wing, J.K. (1968) 'Le évolution et le prognostic de la schizophrénie schizophrénique secondaire', in: *Confrontations psychiatriques: Évolution de la schizophrénie,* pp. 77–87 (Paris: Specia).

Wing, J.K. (1974) Psychiatry in the Soviet Union. *British Medical Journal,* 1, 433–6.

Wing, J.K. (1978) *Schizophrenia: Towards a New Synthesis* (London: Academic Press).

Wing, J.K. (1985) Relapse in schizophrenia. *British Medical Journal (Clinical Research Edition),* 291, 1219–20.

Wizel, M. (1926) Les formes frustes de la schizophénie [The crude forms of schizophrenia]. *Annales Medico-Psychologiques,* 12, 425–51.

Wolman, B.B. (1966) *Vectoriasis praecox of the group of schizophrenias* (Springfield, IL: Charles C. Thomas).

Wolman, B.B. (1973) *Dictionary of Behavioural Science* (London: Van Nostrand Reinhold Company).

World Health Organization (1948) *Manual of the International Statistical Classification of Disease, Injuries and Causes of Death,* 6th ed. (Geneva: World Health Organization).

World Health Organization (1957) *International Classification of Diseases*, 7th ed. (Geneva: World Health Organization).

World Health Organization (1975) *Schizophrenia: A Multinational Study* (Geneva: World Health Organization).

Wyatt, J. and Murphy, D. (1976) Low platelet monoamine oxidase acitvity and schizophrenia. *Schizophrenia Bulletin*, 2, 77–89.

Wyden, P. (1998) *Conquering Schizophrenia: A Father, his Son, and a Medical Breakthrough* (New York: Alfred A. Knopf).

Young, A. (1995) *The Harmony of Illusions, Inventing Post-traumatic Stress Disorder* (Princeton, NJ: Princeton University Press).

Younger, E.G. (1914) *Insanity in Everyday Practice* (London: Baillière, Tindall and Cox).

Zajicek, B. (2014) Nervousness, mild schizophrenia, and the professional jurisdiction of Psychiatry in the USSR, 1918–1936. *Ab Imperio*, 4, 167–94.

Ziehen, T. (1892) Ueber Störungen des Vorstellungsablaufes bei Paranoia. *Archiv für Psychiatrie und Nervenkrankheiten*, 24, 112–54.

Zilboorg, G. (1929) Dynamics of schizophrenic reactions related to childbirth. *American Journal of Psychiatry*, 85, 733–67.

Zilboorg, G. (1931) The deeper layers of schizophrenic psychoses. *American Journal of Psychiatry*, 88, 493–518.

Zilboorg, G. (1941) Ambulatory schizophrenias. *Psychiatry*, 4, 149–55.

Zilboorg, G. (1942) *A History of Medical Psychology* (New York: W.W. Norton).

Zubin, J. (ed.) (1961) *Field Studies in the Mental Disorders* (London: Grune & Stratton).

Index